The Economist

POCKET WORLD IN FIGURES

2021 Edition

Published by
Profile Books Ltd
29 Cloth Fair
London EC1A 7JQ

Published under exclusive licence from
The Economist by Profile Books, 2020

Material researched by
Andrea Burgess, Lisa Davies, Mark Doyle, Ian Emery,
Conrad Heine, Carol Howard, Adam Meara, David McKelvey,
Georgina McKelvey, Christopher Wilson, Pip Wroe

The greatest care has been taken in compiling this book. However,
no responsibility can be accepted by the publishers or compilers
for the accuracy of the information presented.

Typeset in Econ Sans Condensed by MacGuru Ltd

Printed and bound in Italy by L.E.G.O. Spa

A CIP catalogue record for this book is available
from the British Library

ISBN 978 1 78816 497 9

Contents

Introduction

This 2021 edition of *The Economist Pocket World in Figures* presents and analyses data about the world in two sections:

The **world rankings** consider and rank the performance of 188 countries against a range of indicators in five sections: geography and demographics, business and economics, politics and society, health and welfare, and culture and entertainment. The countries included are those which had (in 2018) a population of at least 1m or a GDP of at least $3bn; they are listed on pages 250–53. New rankings this year include topics as diverse as longest coastlines, worst city infrastructure, refugees as a proportion of foreign-born citizens, women on company boards, brain gain, open defecation, handwashing, death due to poor sanitation, ischaemic heart disease, deaths from pollution and malnutrition in children. Some of the rankings data are shown as charts and graphs.

The **country profiles** look in detail at 64 major countries, listed on page 109, plus profiles of the euro area and the world.

Test your *Pocket World in Figures* knowledge with our **World Rankings Quiz** on pages 242–7. Answers can be found in the corresponding world rankings section.

Notes

The extent and quality of the statistics available vary from country to country. Every care has been taken to specify the broad definitions on which the data are based and to indicate cases where data quality or technical difficulties are such that interpretation of the figures is likely to be seriously affected. Nevertheless, figures from individual countries may differ from standard international statistical definitions. The term "country" can also refer to territories or economic entities.

Definitions of the statistics shown are given on the relevant page or in the glossary on pages 248–9. Figures may not add exactly to totals, or percentages to 100, because of rounding or, in the case of GDP, statistical adjustment. Sums of money have generally been converted to US dollars at the official exchange rate ruling at the time to which the figures refer.

Some country definitions

Macedonia was officially known as the Former Yugoslav Republic of Macedonia until February 2019 when it changed to North Macedonia. Data for Cyprus normally refer to Greek Cyprus only. Data for China do not include Hong Kong or Macau. For countries such as Morocco they exclude disputed areas. Congo-Kinshasa refers to the Democratic Republic of Congo, formerly known as Zaire. Congo-Brazzaville refers to the other Congo. Swaziland was officially changed to the Kingdom of Eswatini in April 2018. Sources use both names, but Eswatini is used throughout this book for consistency. Euro area data normally refer to the 19 members that had adopted the euro as at December 31 2018: Austria, Belgium, Cyprus, Estonia, Finland, France, Germany, Greece, Ireland, Italy, Latvia, Lithuania, Luxembourg, Malta, Netherlands, Portugal, Slovakia, Slovenia and Spain. Data referring to the European Union include the United Kingdom, which in June 2016 voted in a referendum to leave the EU. The United Kingdom officially left the EU on January 31st 2020. For more information about the EU, euro area and OECD see the glossary on pages 248–9.

Statistical basis

The all-important factor in a book of this kind is to be able to make reliable comparisons between countries. Although this is never quite possible for the reasons stated above, the best route, which this book takes, is to compare data for the same year or period and to use actual, not estimated, figures wherever possible. In some cases, only OECD members are considered. Where a country's data are excessively out of date, they are excluded. The research for this edition was carried out in 2020 using the latest available sources that present data on an internationally comparable basis.

Data in the country profiles, unless otherwise indicated, refer to the year ending December 31 2018. Life expectancy, crude birth, death and fertility rates are based on 2020–25 estimated averages; energy data are for 2017 and religion data for 2010; marriage and divorce, employment, health and education, consumer goods and services data refer to the latest year for which figures are available.

Other definitions

Data shown in country profiles may not always be consistent with those shown in the world rankings because the definitions or years covered can differ.

Statistics for principal exports and principal imports are normally based on customs statistics. These are generally compiled on different definitions to the visible exports and imports figures shown in the balance of payments section.

Energy-consumption data are not always reliable, particularly for the major oil-producing countries; consumption per person data may therefore be higher than in reality. Energy exports can exceed production and imports can exceed consumption if transit operations distort trade data or oil is imported for refining and re-exported.

Abbreviations and conventions
(see also glossary on pages 248–9)

bn	billion (one thousand million)	km	kilometre
		m	million
EU	European Union	PPP	purchasing power parity
GDP	gross domestic product	TOE	tonnes of oil equivalent
GNI	gross national income	trn	trillion (one thousand billion)
ha	hectare		
kg	kilogram	...	not available

World
rankings

Countries: natural facts

Countries: the largest[a]
'000 sq km

1	Russia	17,098	33	Namibia	824
2	Canada	9,985	34	Pakistan	796
3	United States	9,832	35	Mozambique	786
4	China	9,563	36	Turkey	785
5	Brazil	8,516	37	Chile	757
6	Australia	7,741	38	Zambia	753
7	India	3,287	39	Myanmar	677
8	Argentina	2,780	40	South Sudan	659
9	Kazakhstan	2,725	41	Afghanistan	653
10	Algeria	2,382	42	Somalia	638
11	Congo-Kinshasa	2,345	43	Central African Rep.	623
12	Saudi Arabia	2,150	44	Ukraine	604
13	Mexico	1,964	45	Madagascar	587
14	Indonesia	1,914	46	Botswana	582
15	Sudan	1,879	47	Kenya	580
16	Libya	1,760	48	France	549
17	Iran	1,745	49	Yemen	528
18	Mongolia	1,564	50	Thailand	513
19	Peru	1,285	51	Spain	506
20	Chad	1,284	52	Turkmenistan	488
21	Niger	1,267	53	Cameroon	475
22	Angola	1,247	54	Papua New Guinea	463
23	Mali	1,240	55	Morocco	447
24	South Africa	1,219		Sweden	447
25	Colombia	1,142		Uzbekistan	447
26	Ethiopia	1,104	58	Iraq	435
27	Bolivia	1,099	59	Paraguay	407
28	Mauritania	1,031	62	Zimbabwe	391
29	Egypt	1,001	63	Norway	385
30	Tanzania	947	64	Japan	378
31	Nigeria	924	65	Germany	358
32	Venezuela	912	66	Congo-Brazzaville	342

Coastlines: the longest
Length, km

1	Canada	202,080	15	Denmark	7,314
2	Russia	37,653	16	Turkey	7,200
3	Philippines	36,289	17	India	7,000
4	Japan	29,751	18	Chile	6,435
5	Australia	28,855	19	Croatia	5,835
6	Norway	28,735	20	Papua New Guinea	5,152
7	United States	19,924	21	Argentina	4,989
8	New Zealand	15,134	22	Iceland	4,970
9	China	14,500	23	Spain	4,964
10	Greece	13,676	24	France	4,853
11	United Kingdom	12,429	25	Madagascar	4,828
12	Mexico	9,330	26	Malaysia	4,675
13	Italy	7,600	27	Estonia	3,794
14	Brazil	7,491	28	Cuba	3,735

a Includes freshwater.

Mountains: the highest[a]

		Location	Height (m)
1	Everest	China-Nepal	8,848
2	K2 (Godwin Austen)	China-Pakistan	8,611
3	Kangchenjunga	India-Nepal	8,586
4	Lhotse	China-Nepal	8,516
5	Makalu	China-Nepal	8,463
6	Cho Oyu	China-Nepal	8,201
7	Dhaulagiri	Nepal	8,167
8	Manaslu	Nepal	8,163
9	Nanga Parbat	Pakistan	8,126
10	Annapurna I	Nepal	8,091

Rivers: the longest

		Location	Length (km)
1	Nile	Africa	6,695
2	Amazon	South America	6,516
3	Yangtze (Chang Jiang)	Asia	6,380
4	Mississippi-Missouri system	North America	5,969
5	Ob'-Irtysh	Asia	5,568
6	Yenisey-Angara-Selanga	Asia	5,550
7	Yellow (Huang He)	Asia	5,464
8	Congo	Africa	4,667

Deserts: the largest non-polar

		Location	Area ('000 sq km)
1	Sahara	Northern Africa	8,600
2	Arabian	South-western Asia	2,300
3	Gobi	Mongolia/China	1,300
4	Patagonian	Argentina	673
5	Syrian	Middle East	520
6	Great Basin	South-western United States	490
7	Great Victoria	Western & Southern Australia	419
8	Great Sandy	Western Australia	395

Lakes: the largest

		Location	Area ('000 sq km)
1	Caspian Sea	Central Asia	371
2	Superior	Canada/United States	82
3	Victoria	East Africa	69
4	Huron	Canada/United States	60
5	Michigan	United States	58
6	Tanganyika	East Africa	33
7	Baikal	Russia	31
	Great Bear	Canada	31
9	Malawi	East Africa	30

a Includes separate peaks which are part of the same massif.
Notes: Estimates of the lengths of rivers vary widely depending on, eg, the path to take through a delta. The definition of a desert is normally a mean annual precipitation value equal to 250ml or less.

Population: size and growth

Largest populations
m, 2018

#	Country	Pop.		#	Country	Pop.
1	China	1,427.6		37	Poland	37.9
2	India	1,352.6		38	Afghanistan	37.2
3	United States	327.1		39	Canada	37.1
4	Indonesia	267.7		40	Morocco	36.0
5	Pakistan	212.2		41	Saudi Arabia	33.7
6	Brazil	209.5		42	Uzbekistan	32.5
7	Nigeria	195.9		43	Peru	32.0
8	Bangladesh	161.4		44	Malaysia	31.5
9	Russia	145.7		45	Angola	30.8
10	Japan	127.2		46	Ghana	29.8
11	Mexico	126.2		47	Mozambique	29.5
12	Ethiopia	109.2		48	Venezuela	28.9
13	Philippines	106.7		49	Yemen	28.5
14	Egypt	98.4		50	Nepal	28.1
15	Vietnam	95.5		51	Madagascar	26.3
16	Congo-Kinshasa	84.1		52	North Korea	25.6
17	Germany	83.1		53	Cameroon	25.2
18	Turkey	82.3		54	Ivory Coast	25.1
19	Iran	81.8		55	Australia	24.9
20	Thailand	69.4		56	Taiwan	23.7
21	United Kingdom	67.1		57	Niger	22.4
22	France	65.0		58	Sri Lanka	21.2
23	Italy	60.6		59	Burkina Faso	19.8
24	South Africa	57.8		60	Romania	19.5
25	Tanzania	56.3		61	Mali	19.1
26	Myanmar	53.7		62	Chile	18.7
27	Kenya	51.4		63	Kazakhstan	18.3
28	South Korea	51.2		64	Malawi	18.1
29	Colombia	49.7		65	Zambia	17.4
30	Spain	46.7		66	Guatemala	17.2
31	Argentina	44.4		67	Ecuador	17.1
32	Ukraine	44.2			Netherlands	17.1
33	Uganda	42.7		69	Syria	16.9
34	Algeria	42.2		70	Cambodia	16.3
35	Sudan	41.8		71	Senegal	15.9
36	Iraq	38.4		72	Chad	15.5

Largest populations
m, 2030

#	Country	Pop.		#	Country	Pop.
1	India	1,503.6		11	Mexico	140.9
2	China	1,464.3		12	Philippines	123.7
3	United States	349.6		13	Egypt	120.8
4	Indonesia	299.2			Japan	120.8
5	Nigeria	263.0		15	Congo-Kinshasa	120.0
	Pakistan	263.0		16	Vietnam	104.2
7	Brazil	223.9		17	Iran	92.7
8	Bangladesh	179.0		18	Turkey	89.2
9	Ethiopia	144.9		19	Germany	83.1
10	Russia	143.3		20	Tanzania	79.2

Note: Populations include migrant workers.

Fastest-growing populations
Average annual rate of change, 2015–20, %

1	Bahrain	4.3		Congo-Brazzaville	2.6
2	Niger	3.8		Ethiopia	2.6
3	Equatorial Guinea	3.7		Nigeria	2.6
4	Oman	3.6		Rwanda	2.6
	Uganda	3.6	31	Afghanistan	2.5
6	Maldives	3.5		Guinea-Bissau	2.5
7	Angola	3.3		Iraq	2.5
8	Burundi	3.2		Ivory Coast	2.5
	Congo-Kinshasa	3.2		Liberia	2.5
10	Chad	3.0		Togo	2.5
	Mali	3.0	37	Sudan	2.4
	Tanzania	3.0		Tajikistan	2.4
13	Burkina Faso	2.9		West Bank & Gaza	2.4
	Gambia, The	2.9		Yemen	2.4
	Mozambique	2.9	41	Kenya	2.3
	Zambia	2.9		Qatar	2.3
17	Guinea	2.8	43	Ghana	2.2
	Mauritania	2.8		Kuwait	2.2
	Senegal	2.8	45	Botswana	2.1
	Somalia	2.8		Sierra Leone	2.1
21	Benin	2.7	47	Algeria	2.0
	French Guiana	2.7		Egypt	2.0
	Gabon	2.7		Guatemala	2.0
	Madagascar	2.7		Luxembourg	2.0
	Malawi	2.7		Pakistan	2.0
26	Cameroon	2.6		Papua New Guinea	2.0

Slowest-growing populations
Average annual rate of change, 2015–20, %

1	Puerto Rico	-3.3		Poland	-0.1
2	Lithuania	-1.5		Virgin Islands (US)	-0.1
3	Latvia	-1.2	24	Belarus	0.0
4	Venezuela	-1.1		Cuba	0.0
5	Bosnia & Herz.	-0.9		Guadeloupe	0.0
6	Bulgaria	-0.7		Italy	0.0
	Romania	-0.7		Macedonia	0.0
8	Croatia	-0.6		Montenegro	0.0
	Syria	-0.6		Spain	0.0
10	Bermuda	-0.5	31	Barbados	0.1
	Greece	-0.5		Russia	0.1
	Ukraine	-0.5		Slovakia	0.1
13	Portugal	-0.3		Slovenia	0.1
	Serbia	-0.3	35	Czech Republic	0.2
15	Andorra	-0.2		Estonia	0.2
	Georgia	-0.2		Finland	0.2
	Hungary	-0.2		Mauritius	0.2
	Japan	-0.2		Netherlands	0.2
	Martinique	-0.2		South Korea	0.2
	Moldova	-0.2		Taiwan	0.2
21	Albania	-0.1			

Population: matters of breeding and sex

Crude birth-rates

Births per 1,000 population, 2015–20

Highest			Lowest		
1	Niger	46.3	1	Monaco	6.5
2	Chad	42.4	2	Puerto Rico	7.4
3	Somalia	41.9		South Korea	7.4
4	Mali	41.8	4	Andorra	7.5
5	Congo-Kinshasa	41.4		Japan	7.5
6	Angola	40.9	6	Italy	7.6
7	Burundi	39.3	7	Greece	7.8
8	Gambia, The	38.8		Portugal	7.8
9	Uganda	38.4	9	Bosnia & Herz.	8.2
10	Burkina Faso	38.2	10	Taiwan	8.4
11	Nigeria	38.1	11	Spain	8.5
12	Mozambique	37.7	12	Singapore	8.8
13	Tanzania	36.9	13	Croatia	8.9
14	Guinea	36.6	14	Bulgaria	9.0
15	Benin	36.4	15	Finland	9.4
16	Zambia	36.3		Germany	9.4
17	Ivory Coast	35.9	17	Hungary	9.5
18	Cameroon	35.6	18	Serbia	9.6
19	Central African Rep.	35.4		Ukraine	9.6
	Guinea-Bissau	35.4	20	Qatar	9.7
21	South Sudan	35.2		Slovenia	9.7
22	Senegal	34.7	22	Channel Islands	9.8
23	Malawi	34.3		Malta	9.8
24	Mauritania	33.9		Romania	9.8

Teenage birth rates

Number of births per 1,000 women aged 15–19, 2015–20

1	Niger	186.5	23	Somalia	100.1
2	Mali	169.1	24	Gabon	96.2
3	Chad	161.1	25	Dominican Rep.	94.3
4	Equatorial Guinea	155.6	26	Lesotho	92.7
5	Angola	150.5	27	Togo	89.1
6	Mozambique	148.6	28	Benin	86.1
7	Liberia	136.0		Zimbabwe	86.1
8	Guinea	135.3	30	Venezuela	85.3
9	Malawi	132.7	31	Nicaragua	85.0
10	Central African Rep.	129.1	32	Bangladesh	83.0
11	Congo-Kinshasa	124.2	33	Panama	81.8
12	Zambia	120.1	34	Ecuador	79.3
13	Uganda	118.8	35	Gambia, The	78.2
14	Tanzania	118.4	36	Eswatini	76.7
15	Ivory Coast	117.6	37	Kenya	75.1
16	Sierra Leone	112.8	38	Guyana	74.4
17	Congo-Brazzaville	112.2	39	Honduras	72.9
18	Madagascar	109.6	40	Senegal	72.7
19	Nigeria	107.3	41	Iraq	71.7
20	Cameroon	105.8	42	Mauritania	71.0
21	Guinea-Bissau	104.8	43	Guatemala	70.9
22	Burkina Faso	104.3	44	Paraguay	70.5

Fertility rates
Number of children per woman, 2015–20

Highest			Lowest		
1	Niger	7.0	**1**	South Korea	1.1
2	Somalia	6.1	**2**	Macau	1.2
3	Congo-Kinshasa	6.0		Puerto Rico	1.2
4	Mali	5.9		Singapore	1.2
5	Chad	5.8		Taiwan	1.2
6	Angola	5.6	**6**	Bosnia & Herz.	1.3
7	Burundi	5.5		Cyprus	1.3
8	Nigeria	5.4		Greece	1.3
9	Gambia, The	5.3		Hong Kong	1.3
10	Burkina Faso	5.2		Italy	1.3
11	Uganda	5.0		Moldova	1.3
12	Benin	4.9		Portugal	1.3
	Mozambique	4.9		Spain	1.3
	Tanzania	4.9	**14**	Andorra	1.4
15	Central African Rep.	4.8		Japan	1.4
16	Guinea	4.7		Mauritius	1.4
	Ivory Coast	4.7		Poland	1.4
	Senegal	4.7		Ukraine	1.4
	South Sudan	4.7		United Arab Emirates	1.4
	Zambia	4.7			
21	Afghanistan	4.6			
	Cameroon	4.6			
	Equatorial Guinea	4.6			
	Mauritania	4.6			
25	Congo-Brazzaville	4.5			
	Guinea-Bissau	4.5			

Women[a] who use modern methods of contraception
2018, %

Highest			Lowest		
1	China	82.3	**1**	Chad	6.3
2	United Kingdom	78.8	**2**	South Sudan	6.5
3	Finland	78.3	**3**	Guinea	8.4
4	Nicaragua	77.1	**4**	Congo-Kinshasa	11.4
5	Costa Rica	76.9	**5**	Gambia, The	11.7
6	Brazil	76.8	**6**	Eritrea	13.0
7	Uruguay	76.4	**7**	Benin	14.2
8	Thailand	75.7	**8**	Equatorial Guinea	14.3
9	Colombia	75.3	**9**	Nigeria	14.5
10	France	73.6	**10**	Angola	15.0
11	Hong Kong	72.4	**11**	Mali	15.1
12	Chile	72.1	**12**	Sudan	15.3
	Cuba	72.1	**13**	Ivory Coast	17.0
14	Belgium	71.1	**14**	Guinea-Bissau	17.6
	Ecuador	71.1	**15**	Mauritania	17.7
16	Czech Republic	71.0	**16**	Niger	18.2
17	North Korea	70.9	**17**	Sierra Leone	18.4

a Married women aged 15–49; excludes traditional methods of contraception, such as the rhythm method.

Population: age

Median age[a]

Highest, 2018

1	Monaco	53.8
2	Japan	47.6
3	Italy	46.5
4	Martinique	46.2
5	Germany	45.8
6	Portugal	45.3
7	Andorra	44.9
8	Greece	44.7
9	Isle of Man	44.4
10	Hong Kong	44.2
11	Bulgaria	44.1
	Lithuania	44.1
13	Spain	43.9
	Slovenia	43.9
15	Croatia	43.8
16	Bermuda	43.5
17	Austria	43.4
	Latvia	43.4
	Liechtenstein	43.4
20	Guadeloupe	43.0
21	Finland	42.9
22	Netherlands	42.8

Lowest, 2018

1	Niger	15.1
2	Mali	16.2
3	Chad	16.4
	Uganda	16.4
5	Somalia	16.5
6	Angola	16.6
7	Congo-Kinshasa	16.9
8	Burundi	17.2
	Zambia	17.2
10	Burkina Faso	17.4
	Central African Rep.	17.4
	Mozambique	17.4
13	Guinea	17.6
14	Gambia, The	17.7
	Malawi	17.7
16	Tanzania	17.8
17	Afghanistan	17.9
18	Nigeria	18.0
19	Senegal	18.3
20	Cameroon	18.5

Most old people

% of population aged 65 or over, 2020

1	Monaco	35.1
2	Japan	28.4
3	Italy	23.3
4	Portugal	22.8
5	Finland	22.6
6	Greece	22.3
7	Germany	21.7
	Martinique	21.7
9	Bulgaria	21.5
10	Croatia	21.3
	Malta	21.3
12	Isle of Man	21.1
13	France	20.8
	Puerto Rico	20.8
15	Latvia	20.7
	Slovenia	20.7
17	Lithuania	20.6
18	Virgin Islands (US)	20.5
19	Estonia	20.4
20	Sweden	20.3
21	Denmark	20.2
	Hungary	20.2
23	Czech Republic	20.1

Most young people

% of population aged 0–19, 2020

1	Niger	60.6
2	Mali	58.1
3	Chad	57.8
4	Somalia	57.6
5	Uganda	57.5
6	Angola	57.1
7	Congo-Kinshasa	56.4
8	Central African Rep.	55.9
9	Zambia	55.6
10	Burundi	55.5
11	Burkina Faso	55.4
	Mozambique	55.4
13	Gambia, The	54.6
	Guinea	54.6
15	Malawi	54.4
16	Tanzania	54.3
17	Nigeria	54.1
18	Afghanistan	53.7
19	Senegal	53.1
20	Zimbabwe	52.9
21	Cameroon	52.8
22	Benin	52.6
23	Guinea-Bissau	52.5
	Ivory Coast	52.5

a Age at which there is an equal number of people above and below.

City living

Biggest cities[a]
Population, m, 2018

1	Tokyo, Japan	37.5	26	Lahore, Pakistan	11.7
2	Delhi, India	28.5	27	Bangalore, India	11.4
3	Shanghai, China	25.6	28	Paris, France	10.9
4	São Paulo, Brazil	21.7	29	Bogotá, Colombia	10.6
5	Mexico City, Mexico	21.6	30	Chennai, India	10.5
6	Cairo, Egypt	20.1		Jakarta, Indonesia	10.5
7	Mumbai, India	20.0	32	Lima, Peru	10.4
8	Beijing, China	19.6	33	Bangkok, Thailand	10.2
	Dhaka, Bangladesh	19.6	34	Seoul, South Korea	10.0
10	Osaka, Japan	19.3	35	Hyderabad, India	9.5
11	New York, United States	18.8		Nagoya, Japan	9.5
12	Karachi, Pakistan	15.4	37	London, United Kingdom	9.0
13	Buenos Aires, Argentina	15.0	38	Chicago, United States	8.9
14	Chongqing, China	14.8		Tehran, Iran	8.9
	Istanbul, Turkey	14.8	40	Chengdu, China	8.8
16	Kolkata, India	14.7	41	Taipei, Taiwan	8.5
17	Lagos, Nigeria	13.5	42	Nanjing, China	8.2
	Manila, Philippines	13.5		Wuhan, China	8.2
19	Rio de Janeiro, Brazil	13.3	44	Ho Chi Minh City, Vietnam	8.1
20	Kinshasa, Congo-Kinshasa	13.2	45	Luanda, Angola	7.8
	Tianjin, China	13.2	46	Ahmedabad, India	7.7
22	Guangzhou, China	12.6	47	Kuala Lumpur, Malaysia	7.6
23	Los Angeles, United States	12.5	48	Dongguan, China	7.4
24	Moscow, Russia	12.4		Hong Kong	7.4
25	Shenzhen, China	11.9		Xi'an, China	7.4

City growth[b]
Total % change, 2020–25

	Fastest			Slowest	
1	Bujumbura, Burundi	33.3	1	Bucharest, Romania	-2.4
2	Abomey-Calavi, Benin	30.2	2	Beirut, Lebanon	-1.9
3	Nnewi, Nigeria	29.6	3	Osaka, Japan	-1.3
4	Kampala, Uganda	29.3		Volgograd, Russia	-1.3
5	Mwanza, Tanzania	29.2	5	Hiroshima, Japan	-1.2
6	Abuja, Nigeria	28.4	6	Kitakyushu, Japan	-1.1
7	Uyo, Nigeria	28.3	7	Tokyo, Japan	-1.0
8	Malappuram, India	28.1	8	Kharkiv, Ukraine	-0.9
9	Dar es Salaam, Tanzania	27.8	9	Daegu, South Korea	-0.8
10	Aleppo, Syria	27.2		Samara, Russia	-0.8
11	Bukavu, Congo-Kinshasa	27.0	11	Nizhny Novgorod, Russia	-0.6
12	Ouagadougou, Burkina Faso	26.7		San Juan, Puerto Rico	-0.6
				Sapporo, Japan	-0.6
13	Liuyang, China	26.3	14	Nagoya, Japan	-0.2
14	Port Harcourt, Nigeria	25.6		Naples, Italy	-0.2
15	Antananarivo, Madagascar	25.5		Odessa, Ukraine	-0.2
16	Lusaka, Zambia	25.1	17	Detroit, United States	-0.1
	Mbuji-Mayi, Congo-Kinshasa	25.1		Hamburg, Germany	-0.1
				Omsk, Russia	-0.1

a Urban agglomerations. Data can change from year to year based on reassessments of agglomeration boundaries.
b Urban agglomerations with a population of at least 1m in 2020.

Worst city infrastructure

100=ideal, 0=intolerable, 2018

1	Dhaka, Bangladesh	26.8
2	Algiers, Algeria	30.4
3	Damascus, Syria	32.1
4	Harare, Zimbabwe	35.7
5	Dakar, Senegal	37.5
6	Tehran, Iran	39.3
7	Kathmandu, Nepal	41.1
	Tripoli, Libya	41.1
9	Douala, Cameroon	42.9
	Nairobi, Kenya	42.9
11	Kiev, Ukraine	46.4
	Lagos, Nigeria	46.4
	Mexico City, Mexico	46.4
	Port Moresby, Papua New Guinea	46.4
15	Tashkent, Uzbekistan	48.2

16	Baku, Azerbaijan	50.0
	Phnom Penh, Cambodia	50.0
18	Colombo, Sri Lanka	51.8
	Ho Chi Minh City, Vietnam	51.8
	Karachi, Pakistan	51.8
	Mumbai, India	51.8
22	Abidjan, Ivory Coast	53.6
	Cairo, Egypt	53.6
	Caracas, Venezuela	53.6
	Guatemala City, Guatemala	53.6
26	Al Khobar, Saudi Arabia	55.4
	Hanoi, Vietnam	55.4
	Lusaka, Zambia	55.4
29	Jakarta, Indonesia	57.1

Tallest buildings[a]

Height, metres, 2019

	0	200	400	600	800

Burj Khalifa, Dubai
Shanghai Tower, Shanghai
Makkah Royal Clock Tower, Mecca
Ping An Finance Centre, Shenzhen
Lotte World Tower, Seoul
One World Trade Centre, New York
CTF Finance Centre, Guangzhou
CTF Finance Centre, Tianjin
CITIC Tower, Beijing
Taipei 101, Taipei
World Financial Centre, Shanghai
Int. Commerce Centre, Hong Kong
Lakhta Centre, St Petersburg
Vincom Landmark, Ho Chi Minh City
IFS Tower T1, Changsha
Petronas Twin Towers, Kuala Lumpur
Suzhou IFS, Suzhou
Zifeng Tower, Nanjing
The Exchange 106, Kuala Lumpur
Willis Tower, Chicago
KK100, Shenzhen
Int. Finance Centre, Guangzhou
Centre Tower, Wuhan
432 Park Avenue, New York

a Completed.

Foreign-born and refugees

Largest foreign-born populations[a]

2019, m

1	United States	50.7	24	Switzerland	2.6
2	Germany	13.1	25	Ivory Coast	2.5
	Saudi Arabia	13.1		Japan	2.5
4	Russia	11.6	27	Netherlands	2.3
5	United Kingdom	9.6		Oman	2.3
6	United Arab Emirates	8.6	29	Argentina	2.2
7	France	8.3		Bangladesh	2.2
8	Canada	8.0		Qatar	2.2
9	Australia	7.5		Singapore	2.2
10	Italy	6.3	33	Belgium	2.0
11	Spain	6.1		Israel	2.0
12	Turkey	5.9		Sweden	2.0
13	India	5.2	36	Lebanon	1.9
14	Ukraine	5.0	37	Austria	1.8
15	South Africa	4.2	38	Uganda	1.7
16	Kazakhstan	3.7	39	Venezuela	1.4
17	Thailand	3.6	40	Ethiopia	1.3
18	Malaysia	3.4		Nigeria	1.3
19	Jordan	3.3	42	Greece	1.2
	Pakistan	3.3		South Korea	1.2
21	Kuwait	3.0		Sudan	1.2
22	Hong Kong	2.9		Uzbekistan	1.2
23	Iran	2.7			

Foreign-born as % of total population

2019

Highest			Lowest		
1	United Arab Emirates	87.9	1	Cuba	0.0
2	Qatar	78.7	2	China	0.1
3	Kuwait	72.1		Indonesia	0.1
4	Monaco	68.0		Madagascar	0.1
5	Liechtenstein	67.0		Myanmar	0.1
6	Macau	62.4		Vietnam	0.1
7	Andorra	58.5	7	Haiti	0.2
8	Virgin Islands (US)	54.3		North Korea	0.2
9	Isle of Man	50.7		Philippines	0.2
10	Channel Islands	48.7		Sri Lanka	0.2
11	Guam	47.7	11	Lesotho	0.3
12	Luxembourg	47.4		Morocco	0.3
13	Oman	46.0		Somalia	0.3
14	Bahrain	45.2	14	Afghanistan	0.4
15	Cayman Islands	44.6		Brazil	0.4
16	French Guiana	40.4		Honduras	0.4
17	Hong Kong	39.6		India	0.4
18	Saudi Arabia	38.3		Papua New Guinea	0.4
19	Singapore	37.1	19	Cambodia	0.5
20	Jordan	33.1		Egypt	0.5
21	Bermuda	30.9		Eritrea	0.5
22	Australia	30.0		Guatemala	0.5
23	Switzerland	29.9		Tunisia	0.5

a Includes migrant workers and illegal immigrants.

Biggest increases in foreign-born populations[a]
Average annual % change, 2015–19

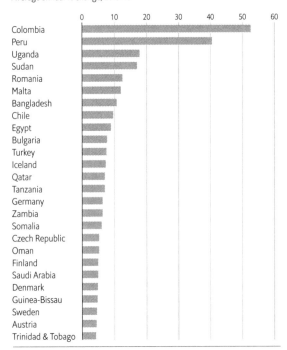

	0	10	20	30	40	50	60
Colombia							
Peru							
Uganda							
Sudan							
Romania							
Malta							
Bangladesh							
Chile							
Egypt							
Bulgaria							
Turkey							
Iceland							
Qatar							
Tanzania							
Germany							
Zambia							
Somalia							
Czech Republic							
Oman							
Finland							
Saudi Arabia							
Denmark							
Guinea-Bissau							
Sweden							
Austria							
Trinidad & Tobago							

Refugees and asylum-seekers
As % of foreign-born population, 2019

1	Jordan	87.5		8	Ethiopia	71.2
2	Lebanon	83.6		9	Tanzania	69.3
3	Chad	80.5		10	Cameroon	67.9
4	Uganda	80.4			Syria	67.9
5	Iraq	79.0		12	Turkey	64.4
6	Sudan	75.6		13	Egypt	57.4
7	Yemen	72.8		14	Niger	56.4

Stock of a population displaced by conflict
'000, as of end 2018

1	Syria	6,119		8	Ethiopia	2,137
2	Colombia	5,761		9	Sudan	2,072
3	Congo-Kinshasa	3,081		10	Iraq	1,962
4	Somalia	2,648		11	South Sudan	1,869
5	Afghanistan	2,598		12	Turkey	1,097
6	Yemen	2,324		13	Ukraine	800
7	Nigeria	2,216		14	Cameroon	668

a Includes migrants and illegal immigrants.

Refugees,[a] country of origin

'000, 2018

1	Syria	6,654.4		11	Iraq	372.3
2	Afghanistan	2,681.3		12	Vietnam	334.5
3	South Sudan	2,285.3		13	Nigeria	276.9
4	Myanmar	1,145.2		14	Rwanda	247.5
5	Somalia	949.7		15	China	212.1
6	Sudan	724.8		16	Mali	158.3
7	Congo-Kinshasa	720.3		17	Colombia	138.6
8	Central African Rep.	590.9		18	Pakistan	132.3
9	Eritrea	507.3		19	Iran	129.9
10	Burundi	387.9		20	Sri Lanka	114.0

Countries with largest refugee[a] populations

'000, 2018

1	Turkey	3,681.7		11	Congo-Kinshasa	529.1
2	Pakistan	1,404.0		12	Chad	451.2
3	Uganda	1,165.7		13	Kenya	421.2
4	Sudan	1,078.3		14	Cameroon	380.3
5	Germany	1,063.8		15	France	368.4
6	Iran	979.4		16	China	321.8
7	Lebanon	949.7		17	United States	313.2
8	Bangladesh	906.6		18	South Sudan	291.8
9	Ethiopia	903.2		19	Iraq	283.0
10	Jordan	715.3		20	Tanzania	278.3

Applications for asylum by country of origin

'000, 2018

1	Venezuela	348.9		11	Honduras	43.4
2	Afghanistan	139.0		12	Turkey	39.1
3	Syria	126.4		13	Guatemala	38.5
4	Iraq	99.8		14	China	38.1
5	Congo-Kinshasa	67.5		15	Sudan	38.0
6	Nigeria	59.7		16	Somalia	37.5
7	Pakistan	54.5		17	Mexico	35.1
8	Iran	51.5		18	Albania	33.3
9	El Salvador	50.0		19	Russia	32.5
10	Eritrea	47.4		20	Nicaragua	31.7

Countries where asylum applications were lodged

'000, 2018

1	Germany	319.1		11	Italy	48.5
2	United States	309.1		12	Sweden	44.6
3	Peru	192.5		13	Australia	42.7
4	France	182.3		14	Mexico	29.6
5	Turkey	84.2		15	Malaysia	29.4
6	Greece	82.3		16	Costa Rica	29.0
7	Brazil	80.0		17	Egypt	26.0
8	Canada	62.6		18	Netherlands	24.0
9	Spain	55.7		19	Belgium	22.6
10	United Kingdom	52.6		20	Uganda	20.1

a   According to UNHCR. Includes people in refugee-like situations.

The world economy

Biggest economies

GDP, $bn, 2018

1	United States	20,544	24	Belgium	543
2	China	13,608	25	Argentina	520
3	Japan	4,971	26	Thailand	505
4	Germany	3,948	27	Austria	455
5	United Kingdom	2,855	28	Iran	446
6	France[a]	2,778	29	Norway	434
7	India	2,719	30	United Arab Emirates	414
8	Italy	2,084	31	Nigeria	397
9	Brazil	1,869	32	Ireland	382
10	Canada	1,713	33	Israel	371
11	Russia	1,658	34	South Africa	368
12	South Korea	1,619	35	Singapore	364
13	Australia	1,434	36	Hong Kong	363
14	Spain	1,419	37	Malaysia	359
15	Mexico	1,221	38	Denmark	356
16	Indonesia	1,042	39	Colombia	331
17	Netherlands	914		Philippines	331
18	Saudi Arabia	787	41	Pakistan	315
19	Turkey	771	42	Chile	298
20	Switzerland	705	43	Finland	277
21	Taiwan	613	44	Bangladesh	274
22	Poland	586	45	Egypt	251
23	Sweden	556			

Biggest economies by purchasing power

GDP PPP, $bn, 2018

1	China	25,399	24	Pakistan	1,181
2	United States	20,544	25	Nigeria	1,173
3	India	10,500	26	Malaysia	1,002
4	Japan	5,415	27	Netherlands	971
5	Germany	4,401	28	Philippines	955
6	Russia	4,051	29	Argentina	917
7	Indonesia	3,501	30	South Africa	791
8	Brazil	3,372	31	Colombia	745
9	United Kingdom	3,057	32	United Arab Emirates	723
10	France	3,037	33	Vietnam	712
11	Italy	2,528	34	Bangladesh	705
12	Mexico	2,504	35	Iraq	670
13	Turkey	2,311	36	Algeria	654
14	South Korea	2,071	37	Belgium	587
15	Saudi Arabia	1,865	38	Switzerland	580
16	Spain	1,856	39	Singapore	573
17	Canada	1,784	40	Romania	549
18	Iran	1,596	41	Sweden	542
19	Thailand	1,323	42	Kazakhstan	510
20	Australia	1,291	43	Austria	491
21	Taiwan	1,252	44	Hong Kong	481
22	Egypt	1,222	45	Chile	472
23	Poland	1,190	46	Peru	461

a Includes overseas territories.   b 2018   c IMF coverage.

Regional GDP

$bn, 2019		*% annual growth 2014–19*	
World	86,599	World	3.5
Advanced economies	51,744	Advanced economies	2.1
Euro area (19)	13,314	Euro area (19)	1.9
G7	39,627	G7	1.9
Emerging & dev. Asia	20,318	Emerging & developing Asia	6.5
Emerging & dev. Europe	3,826	Emerging & developing Europe	2.3
Latin America & the Carib.	5,188	Latin America & the Carib.	0.4
Middle East, N. Africa & Central Asia	3,829	Middle East, N. Africa & Central Asia	2.5
Sub-Saharan Africa	1,694	Sub-Saharan Africa	2.8

Regional purchasing power

GDP, % of total, 2019		*$ per person, 2019*	
World	100.0	World[b]	17,948
Advanced economies	40.3	Advanced economies	53,369
Euro area (19)	11.2	Euro area (19)	46,759
G7	29.7	G7	54,909
Emerging & developing Asia	34.1	Emerging & dev. Asia	13,349
Emerging & dev. Europe	7.1	Emerging & dev. Europe	26,437
Latin America & the Carib.	7.2	Latin America & the Carib.	16,357
Middle East, N. Africa & Central Asia	8.1	Middle East, N. Africa & Central Asia	14,188
Sub-Saharan Africa	3.1	Sub-Saharan Africa	4,195

Regional population

% of total (7.6bn), 2019		*No. of countries[c], 2018*	
World	100.0	World	194
Advanced economies	14.2	Advanced economies	39
Euro area (19)	4.5	Euro area (19)	19
G7	10.2	G7	7
Emerging & developing Asia	48.1	Emerging & developing Asia	30
Emerging & dev. Europe	5.0	Emerging & developing Europe	16
Latin America & the Carib.	8.3	Latin America & the Caribbean	33
Middle East, N. Africa & Central Asia	10.7	Middle East, N. Africa & Central Asia	31
Sub-Saharan Africa	13.7	Sub-Saharan Africa	45

Regional international trade

Exports of goods & services *% of total, 2019*		*Current-account balances* *$bn, 2019*	
World	100.0	World	291
Advanced economies	63.0	Advanced economies	305
Euro area (19)	26.2	Euro area (19)	377
G7	33.4	G7	-182
Emerging & developing Asia	18.1	Emerging & developing Asia	83
Emerging & dev. Europe	6.2	Emerging & developing Europe	60
Latin America & the Caribbean	5.1	Latin America & the Carib.	-81
Middle East, N. Africa & Central Asia	5.9	Middle East, N. Africa & Central Asia	-15
Sub-Saharan Africa	1.7	Sub-Saharan Africa	-62

Living standards

Highest GDP per person
$, 2018

1	Monaco	185,741	31	Japan	39,290
2	Liechtenstein[a]	165,028	32	Virgin Islands (US)[c]	35,938
3	Luxembourg	116,640	33	Guam	35,713
4	Macau	87,209	34	Italy	34,483
5	Bermuda[b]	85,748	35	Kuwait	33,994
6	Switzerland	82,797	36	Bahamas	32,218
7	Norway	81,697	37	Puerto Rico	31,651
8	Cayman Islands[c]	81,125	38	Brunei	31,628
9	Isle of Man[c]	80,989	39	South Korea	31,363
10	Ireland	78,806	40	Spain	30,371
11	Iceland	73,191	41	Malta	30,098
12	Qatar	68,794	42	Cyprus	28,690
13	Singapore	64,582	43	Slovenia	26,124
14	United States	62,795	44	Taiwan	25,008
15	Denmark	61,350	45	Bahrain	24,051
16	Australia	57,374	46	Portugal	23,408
17	Sweden	54,608	47	Saudi Arabia	23,339
18	Netherlands	53,024	48	Estonia	23,266
19	Austria	51,462	49	Czech Republic	23,079
20	Finland	50,152	50	Greece	20,324
21	Hong Kong	48,676	51	Curaçao	19,568
22	Germany	47,603	52	Slovakia	19,443
23	Belgium	47,519	53	Lithuania	19,153
24	Canada	46,233	54	Barbados	17,949
25	United Arab Emirates	43,005	55	Latvia	17,861
26	United Kingdom	42,944	56	Uruguay	17,278
27	Andorra	42,030	57	Trinidad & Tobago	17,130
28	New Zealand	41,945	58	Oman	16,415
29	Israel	41,715	59	Hungary	16,162
30	France	41,464	60	Chile	15,923

Lowest GDP per person
$, 2018

1	Burundi	272	17	Gambia, The	716
2	Somalia	315	18	Chad	728
3	Eritrea	332	19	Ethiopia	772
4	South Sudan	353	20	Rwanda	773
5	Malawi	389	21	Guinea-Bissau	778
6	Niger	414	22	Tajikistan	827
7	Central African Rep.	476	23	Haiti	868
8	Mozambique	499	24	Guinea	879
9	Afghanistan	521	25	Mali	900
10	Madagascar	528	26	Benin	902
11	Sierra Leone	534	27	Yemen	944
12	Congo-Kinshasa	562	28	Sudan	977
13	Uganda	643	29	Nepal	1,034
14	Liberia	677	30	Tanzania	1,061
15	Togo	679	31	Mauritania	1,189
16	Burkina Faso	715	32	Kyrgyzstan	1,281

a 2016   b 2013   c 2017

Highest purchasing power
GDP per person in PPP (US = 100), 2018 or latest

1	Liechtenstein	221.5	35	Italy	66.6
2	Qatar	202.1	36	New Zealand	65.3
3	Macau	197.3	37	South Korea	63.9
4	Monaco	184.3	38	Israel	63.6
5	Luxembourg	180.5	39	Czech Republic	63.3
6	Singapore	161.7	40	Spain	63.2
7	Ireland	132.5	41	Puerto Rico	63.0
8	Brunei	128.9	42	Cyprus	61.3
9	United Arab Emirates	119.6	43	Slovenia	60.6
10	Kuwait	116.1	44	Estonia	57.3
11	Cayman Islands	115.6	45	Lithuania	56.5
12	Switzerland	108.4	46	Slovakia	53.7
13	Norway	104.3	47	Portugal	53.2
14	Hong Kong	102.9	48	Bahamas	51.1
15	United States	100.0	49	Trinidad & Tobago	51.0
16	Iceland	91.3	50	Malaysia	50.6
17	Netherlands	89.7	51	Poland	49.9
18	Denmark	88.7	52	Hungary	49.5
19	Austria	88.3	53	Latvia	48.3
20	Saudi Arabia	88.1	54	Greece	47.1
21	Sweden	84.7	55	Romania	44.9
22	Germany	84.5	56	Turkey	44.7
	Taiwan	84.5	57	Kazakhstan	44.4
24	Bermuda	83.7	58	Curaçao	44.2
25	Australia	82.3	59	Croatia	43.9
26	Belgium	81.9		Russia	43.9
27	Finland	77.1	61	Panama	40.7
28	Canada	76.6	62	Chile	40.2
29	Bahrain	75.3	63	Mauritius	37.8
30	United Kingdom	73.2	64	Uruguay	37.5
31	France	72.2	65	Equatorial Guinea	36.2
32	Japan	68.2	66	Bulgaria	35.0
33	Malta	67.8	67	Libya	33.1
34	Oman	66.7	68	Montenegro	32.9

Lowest purchasing power
GDP per person in PPP (US = 100), 2018 or latest

1	Burundi	1.19	15	Afghanistan	3.11
2	Central African Rep.	1.37	16	Chad	3.13
3	Congo-Kinshasa	1.48	17	Burkina Faso	3.16
4	Eritrea	1.64	18	Ethiopia	3.22
5	Niger	1.69	19	Uganda	3.25
6	Liberia	2.08	20	Rwanda	3.59
7	Malawi	2.09	21	Mali	3.69
8	Mozambique	2.32	22	Benin	3.86
9	South Sudan	2.39	23	Guinea	3.99
10	Sierra Leone	2.55	24	Yemen	4.10
11	Togo	2.82	25	Gambia, The	4.16
12	Guinea-Bissau	2.87	26	Zimbabwe	4.82
13	Haiti	2.97	27	Nepal	4.92
14	Madagascar	3.01	28	Lesotho	5.13

The quality of life

Human development index[a]

Highest, 2018			Lowest, 2018		
1	Norway	95.4	1	Niger	37.7
2	Switzerland	94.6	2	Central African Rep.	38.1
3	Ireland	94.2	3	Chad	40.1
4	Germany	93.9	4	South Sudan	41.3
	Hong Kong	93.9	5	Burundi	42.3
6	Australia	93.8	6	Mali	42.7
	Iceland	93.8	7	Burkina Faso	43.4
8	Sweden	93.7		Eritrea	43.4
9	Singapore	93.5	9	Sierra Leone	43.8
10	Netherlands	93.3	10	Mozambique	44.6
11	Denmark	93.0	11	Congo-Kinshasa	45.9
12	Finland	92.5	12	Guinea-Bissau	46.1
13	Canada	92.2	13	Yemen	46.3
14	New Zealand	92.1	14	Liberia	46.5
15	United Kingdom	92.0	15	Gambia, The	46.6
	United States	92.0		Guinea	46.6
17	Belgium	91.9	17	Ethiopia	47.0
18	Liechtenstein	91.7	18	Malawi	48.5
19	Japan	91.5	19	Afghanistan	49.6
20	Austria	91.4	20	Haiti	50.3
21	Luxembourg	90.9	21	Sudan	50.7
22	Israel	90.6	22	Togo	51.3
	South Korea	90.6	23	Senegal	51.4
24	Slovenia	90.2	24	Ivory Coast	51.6

Gini coefficient[b]

Highest, 2010–17			Lowest, 2010–17		
1	South Africa	63.0	1	Ukraine	25.0
2	Namibia	59.1	2	Belarus	25.4
3	Zambia	57.1		Slovenia	25.4
4	Central African Rep.	56.2	4	Czech Republic	25.9
5	Lesotho	54.2		Moldova	25.9
6	Mozambique	54.0	6	Slovakia	26.5
7	Botswana	53.3	7	Finland	27.1
	Brazil	53.3	8	Kyrgyzstan	27.3
9	Eswatini	51.5	9	Kazakhstan	27.5
10	Guinea-Bissau	50.7		Norway	27.5
11	Honduras	50.5	11	Algeria	27.6
12	Panama	49.9	12	Belgium	27.7
13	Colombia	49.7	13	Iceland	27.8
14	Congo-Brazzaville	48.9	14	Denmark	28.2
15	Paraguay	48.8		Netherlands	28.2
16	Costa Rica	48.3	16	Serbia	28.5
	Guatemala	48.3	17	Timor-Leste	28.7

a GDP or GDP per person is often taken as a measure of how developed a country is, but its usefulness is limited as it refers only to economic welfare. The UN Development Programme combines statistics on average and expected years of schooling and life expectancy with income levels (now GNI per person, valued in PPP US$). The HDI is shown here scaled from 0 to 100; countries scoring over 80 are considered to have very high human development, 70–79 high, 55–69 medium and those under 55 low.
b The lower its value, the more equally household income is distributed.

Household wealth
$trn, highest, 2018

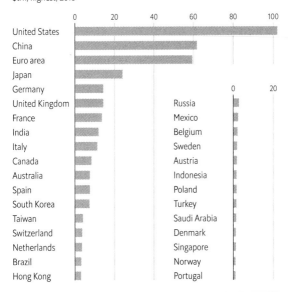

United States
China
Euro area
Japan
Germany
United Kingdom
France
India
Italy
Canada
Australia
Spain
South Korea
Taiwan
Switzerland
Netherlands
Brazil
Hong Kong

Russia
Mexico
Belgium
Sweden
Austria
Indonesia
Poland
Turkey
Saudi Arabia
Denmark
Singapore
Norway
Portugal

Number of millionaires
Persons with net worth above $1m, 2018, m

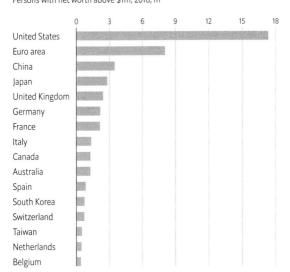

United States
Euro area
China
Japan
United Kingdom
Germany
France
Italy
Canada
Australia
Spain
South Korea
Switzerland
Taiwan
Netherlands
Belgium

Economic growth

Highest economic growth
Average annual % increase in real GDP, 2008–18

1	Ethiopia	9.8		Papua New Guinea	5.6
2	Turkmenistan	8.7		Zambia	5.6
3	China	7.9	28	Niger	5.5
4	Zimbabwe	7.8	29	Burkina Faso	5.4
5	Laos	7.5		Indonesia	5.4
	Myanmar	7.5		Uganda	5.4
7	India	7.1	32	Guinea	5.3
	Rwanda	7.1		Iraq	5.3
9	Mongolia	6.9		Macau	5.3
	Uzbekistan	6.9		Sri Lanka	5.3
11	Tajikistan	6.7	36	Dominican Rep.	5.2
12	Ghana	6.6		Turkey	5.2
13	Qatar	6.5	38	Ireland	5.0
14	Bangladesh	6.4	39	West Bank & Gaza	4.9
15	Afghanistan	6.3	40	Bolivia	4.8
	Cambodia	6.3		Maldives	4.8
	Tanzania	6.3		Malta	4.8
18	Vietnam	6.1		Senegal	4.8
19	Congo-Kinshasa	6.0	44	Malaysia	4.7
	Ivory Coast	6.0	45	Nepal	4.6
21	Mozambique	5.9		Singapore	4.6
	Panama	5.9	47	Malawi	4.5
23	Philippines	5.8	48	Benin	4.4
24	Togo	5.7		Mali	4.4
25	Kenya	5.6		Peru	4.4

Lowest economic growth
Average annual % change in real GDP, 2008–18

1	Venezuela	-6.0	23	Spain	0.4
2	Virgin Islands (US)[a]	-4.2	24	Cayman Islands[a]	0.5
3	Yemen	-3.7	25	Cyprus	0.6
4	Libya	-3.5	26	Guadeloupe[c]	0.7
5	Equatorial Guinea	-2.7		Japan	0.7
	Greece	-2.7		Kuwait	0.7
7	Timor-Leste	-2.1		Martinique[c]	0.7
8	Ukraine	-1.5		Slovenia	0.7
9	Puerto Rico	-1.4	31	Euro area	0.8
10	Andorra	-0.7		Latvia	0.8
	Curaçao	-0.7		Russia	0.8
12	Barbados	-0.6	34	Argentina	0.9
	Central African Rep.	-0.6		France	0.9
	Trinidad & Tobago	-0.6		Guam	0.9
15	Italy	-0.3		Netherlands	0.9
16	Brunei	-0.1	38	Austria	1.0
	Croatia	-0.1	39	Denmark	1.1
18	North Korea[b]	0.0	40	Belgium	1.2
19	Bahamas	0.1		Brazil	1.2
20	Finland	0.2		Norway	1.2
	Jamaica	0.2		Réunion[c]	1.2
	Portugal	0.2		Serbia	1.2

Highest economic growth
Average annual % increase in real GDP, 1998–2008

1	Equatorial Guinea	22.8	11	Chad	8.7
2	Timor-Leste[d]	17.7		Kazakhstan	8.7
3	Azerbaijan	15.2	13	Tajikistan	8.3
4	Qatar[d]	12.6		Turkmenistan	8.3
5	Myanmar	12.4	15	Rwanda	8.2
6	Armenia	10.4	16	Mozambique	8.0
7	China	10.2	17	Ethiopia	7.6
8	Cambodia	9.7		Trinidad & Tobago	7.6
9	Macau	9.0	19	Belarus	7.5
10	Angola	8.8	20	Uganda	7.2

Lowest economic growth
Average annual % change in real GDP, 1998–2008

1	Zimbabwe	-6.7	11	Germany	1.5
2	Gabon	-0.4		Jamaica	1.5
3	Eritrea	0.0	13	Guam[e]	1.6
4	Liberia[d]	0.3		Portugal	1.6
5	Ivory Coast	0.5	15	Denmark	1.8
6	Haiti	0.7		Guyana	1.8
7	Curaçao[d]	0.9		Puerto Rico	1.8
8	Japan	1.0		Togo	1.8
9	Italy	1.2		West Bank & Gaza	1.8
10	Uruguay	1.4			

Highest services growth
Average annual % increase in real terms, 2008–18

1	Liberia	13.2	10	Mongolia[f]	7.2
2	Zimbabwe	12.6		Togo	7.2
3	Ethiopia	11.5	12	Laos[g]	7.0
4	Myanmar[f]	9.1	13	Mozambique	6.8
5	China	8.4		Zambia	6.8
6	India	8.2	15	Azerbaijan	6.7
7	Afghanistan	8.1		Macau	6.7
	Rwanda	8.1		Tanzania[a]	6.7
9	Uzbekistan	7.5	18	Timor-Leste[a]	6.6

Lowest services growth
Average annual % change in real terms, 2008–18

1	Libya	-4.7		Suriname	-0.3
	Syria	-4.7	11	Bermuda	0.0
3	Yemen	-2.7	12	Jamaica	0.1
4	Greece	-2.6	13	Bahamas	0.2
5	Puerto Rico	-2.4		Italy	0.2
6	Ukraine	-1.0	15	Japan[a]	0.3
7	Virgin Islands (US)	-0.9	16	Finland	0.4
8	North Korea	-0.7	17	Cayman Islands[a]	0.5
9	South Sudan[c]	-0.3		Trinidad & Tobago	0.5

a 2008–17 b 2009–18 c 2008–15 d 2000–08 e 2002–08 f 2010–18 g 2012–18
Note: Rankings of highest and lowest industrial growth 2008–18 can be found on page 44.

Trading places

Biggest exporters

% of total world exports (goods, services and income), 2018

1	Euro area (19)	16.49	22	Luxembourg	1.32
2	United States	12.37	23	Australia	1.30
3	China	10.00	24	Poland	1.17
4	Germany	7.40	25	Thailand	1.16
5	Japan	4.25	26	Saudi Arabia	1.15
6	United Kingdom	4.03	27	Hong Kong	1.13
7	France	3.87	28	Sweden	1.06
8	Netherlands	3.81	29	Austria	0.99
9	South Korea	2.64		Brazil	0.99
10	Italy	2.58	31	Malaysia	0.90
11	Canada	2.25		Vietnam	0.90
12	Switzerland	2.21	33	Turkey	0.84
13	Spain	1.96	34	Denmark	0.79
14	Ireland	1.95	35	Indonesia	0.76
15	India	1.93	36	Norway	0.73
	Russia	1.93	37	Czech Republic	0.70
17	Singapore	1.81	38	Hungary	0.52
18	Belgium	1.79	39	Finland	0.44
19	Mexico	1.70	40	Israel	0.43
20	Taiwan	1.53	41	South Africa	0.41
21	United Arab Emirates	1.42			

Trade dependency

Trade[a] as % of GDP, 2018

Most			Least		
1	Vietnam	96.0	1	Cuba[b]	7.1
2	Slovakia	84.5	2	Bermuda	7.7
3	Hungary	67.0	3	Yemen	9.1
4	Slovenia	66.8	4	Ethiopia	9.7
5	United Arab Emirates	66.7	5	United States	10.3
6	Cambodia	64.7	6	Tanzania	10.5
7	Czech Republic	63.8	7	Brazil	11.4
8	Lithuania	63.5	8	Hong Kong	11.5
9	Belgium	59.9	9	Argentina	12.0
10	Puerto Rico	58.5	10	Timor-Leste	12.4
11	Belarus	58.1	11	Kenya	12.8
	Netherlands	58.1	12	Sudan	12.9
13	Lesotho	57.6	13	Nigeria	13.1
14	South Sudan	54.5		Pakistan	13.1
15	Taiwan	54.0	15	Macau	13.8
16	Macedonia	53.5	16	Burundi	14.1
17	Malaysia	53.4		Cameroon	14.1
18	Bulgaria	51.9	18	Colombia	14.2
19	Singapore	51.6	19	Turkmenistan	14.6
20	Estonia	50.8	20	Japan	14.7
21	Suriname	50.5	21	Central African Rep.	15.0

Notes: The figures are drawn wherever possible from balance of payment statistics so have differing definitions from statistics taken from customs or similar sources. For Hong Kong and Singapore, only domestic exports and retained imports are used. Euro area data exclude intra-euro area trade.

a Average of imports plus exports of goods.   b Estimate.

Biggest traders of goods[a]
% of world, 2019

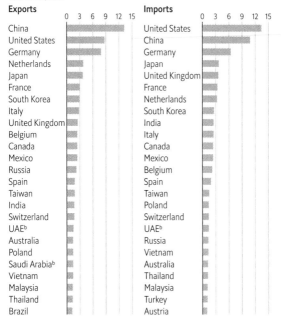

Exports		Imports	
China		United States	
United States		China	
Germany		Germany	
Netherlands		Japan	
Japan		United Kingdom	
France		France	
South Korea		Netherlands	
Italy		South Korea	
United Kingdom		India	
Belgium		Italy	
Canada		Canada	
Mexico		Mexico	
Russia		Belgium	
Spain		Spain	
Taiwan		Taiwan	
India		Poland	
Switzerland		Switzerland	
UAE[b]		UAE[b]	
Australia		Russia	
Poland		Vietnam	
Saudi Arabia[b]		Australia	
Vietnam		Thailand	
Malaysia		Malaysia	
Thailand		Turkey	
Brazil		Austria	

Biggest earners from services and income
% of world exports of services and income, 2018

1	Euro area (19)	18.98	19	South Korea	1.31
2	United States	18.02	20	Sweden	1.22
3	United Kingdom	6.61	21	Australia	1.11
4	Germany	5.82		Russia	1.11
5	Netherlands	4.99	23	Denmark	1.03
6	France	4.81	24	Austria	1.02
7	Japan	4.68	25	United Arab Emirates	0.89
8	China	4.53	26	Norway	0.85
9	Luxembourg	3.39		Taiwan	0.85
10	Hong Kong	3.03	28	Thailand	0.81
11	Singapore	2.98	29	Poland	0.79
12	Ireland	2.93	30	Israel	0.61
13	Switzerland	2.86		Turkey	0.61
14	India	2.13	32	Malaysia	0.50
15	Spain	2.12	33	Finland	0.49
16	Italy	2.04	34	Greece	0.48
17	Canada	1.90		Macau	0.48
18	Belgium	1.83		Philippines	0.48

a Individual countries only. b Estimate.

Balance of payments: current account

Largest surpluses
$m, 2018

1	Euro area (19)	425,196	26	Austria	10,799
2	Germany	293,071	27	Sweden	9,665
3	Japan	176,098	28	Israel	9,277
4	Russia	113,733	29	Venezuela	8,613
5	Netherlands	99,063	30	Puerto Rico	8,148
6	South Korea	77,467	31	Malaysia	7,590
7	Taiwan	71,873	32	Angola	7,403
8	Saudi Arabia	70,606	33	Azerbaijan	6,051
9	Singapore	64,114	34	Vietnam	5,899
10	Switzerland	57,857	35	Nigeria	5,334
11	Italy	51,525	36	Papua New Guinea	5,099
12	Ireland	40,901	37	Bulgaria	3,492
13	United Arab Emirates	40,490	38	Luxembourg	3,333
14	Iraq	34,370	39	Slovenia	3,073
15	Norway	31,138	40	Cuba[a]	2,531
16	Thailand	28,457	41	Malta	1,621
17	Spain	27,308	42	Trinidad & Tobago	1,386
18	Iran	26,741	43	Czech Republic	1,190
19	China	25,499	44	Brunei	1,068
20	Denmark	24,781	45	Croatia	1,028
21	Kuwait	24,049	46	Bermuda	958
22	Macau	19,060	47	Portugal	916
23	Qatar	16,652	48	Lithuania	818
24	Hong Kong	13,514	49	Iceland	792
25	Libya	11,276	50	Estonia	612

Largest deficits
$m, 2018

1	United States	-490,991	22	Bangladesh	-7,593
2	United Kingdom	-110,231	23	Belgium	-7,350
3	India	-65,599	24	Greece	-6,249
4	Canada	-42,993	25	Morocco	-6,205
5	Brazil	-41,540	26	Poland	-5,820
6	Indonesia	-30,633	27	Panama	-5,355
7	Australia	-29,175	28	Sudan	-4,679
8	Argentina	-27,276	29	Ethiopia	-4,611
9	Mexico	-23,004	30	Finland	-4,570
10	Turkey	-20,745	31	Mozambique	-4,501
11	Pakistan	-19,482	32	Tunisia	-4,429
12	France	-19,014	33	Ukraine	-4,367
13	Algeria	-16,700	34	Kenya	-4,362
14	South Africa	-13,384	35	Oman	-4,347
15	Colombia	-13,047	36	Afghanistan	-3,897
16	Lebanon	-12,445	37	Peru	-3,594
17	Romania	-10,503	38	Uzbekistan	-3,594
18	Chile	-9,157	39	Cambodia	-2,992
19	Philippines	-8,729	40	Jordan	-2,850
20	New Zealand	-7,708	41	Sri Lanka	-2,814
21	Egypt	-7,698	42	Nepal	-2,775

Note: Euro area data exclude intra-euro area trade.
a Estimate.

Largest surpluses as % of GDP
$m, 2018

1	Macau	34.6	26	Trinidad & Tobago	5.8
2	Libya	23.3	27	Slovenia	5.7
3	Papua New Guinea	21.7	28	Thailand	5.6
4	Singapore	17.6	29	Bulgaria	5.4
5	Kuwait	17.1	30	South Korea	4.8
6	Iraq	15.3	31	Luxembourg	4.7
7	Bermuda	13.2	32	Hong Kong	3.7
8	Azerbaijan	12.9	33	Japan	3.5
9	Taiwan	12.2	34	Euro area (19)	3.1
10	Malta	11.1		Iceland	3.1
11	Netherlands	10.8	36	Cuba[a]	2.5
12	Ireland	10.7		Israel	2.5
13	United Arab Emirates	9.8		Italy	2.5
14	Saudi Arabia	9.0	39	Austria	2.4
15	Venezuela	8.8		Turkmenistan	2.4
16	Qatar	8.7		Vietnam	2.4
17	Switzerland	8.2	42	Malaysia	2.1
18	Puerto Rico	8.1	43	Estonia	2.0
19	Brunei	7.9	44	Spain	1.9
20	Germany	7.4	45	Botswana	1.8
21	Norway	7.2	46	Croatia	1.7
22	Angola	7.0		Sweden	1.7
	Denmark	7.0	48	Lithuania	1.5
24	Russia	6.9	49	Eswatini	1.3
25	Iran	6.0		Nigeria	1.3

Largest deficits as % of GDP
$m, 2018

1	Mozambique	-30.6	22	Moldova	-10.6
2	Curaçao	-28.7	23	Algeria	-9.6
3	Guyana	-27.6		Nepal	-9.6
4	Maldives	-26.1	25	Armenia	-9.4
5	Lebanon	-22.0	26	Senegal	-9.2
6	Sierra Leone	-20.9	27	Fiji	-8.5
7	Liberia	-20.7	28	Uganda	-8.4
8	Malawi	-20.2	29	Central African Rep.	-8.2
9	Cayman Islands[b]	-19.4		Panama	-8.2
10	Afghanistan	-20.1	31	Laos	-8.0
11	Mauritania	-18.6	32	Rwanda	-7.8
12	Niger	-17.5	33	Haiti	-7.6
13	Montenegro	-17.1		Kosovo	-7.6
14	Mongolia	-14.6	35	Timor-Leste	-7.4
15	Cambodia	-12.2	36	Georgia	-7.2
16	Bahamas	-12.1	37	Uzbekistan	-7.1
17	Burundi	-11.9	38	South Sudan	-6.9
	Kyrgyzstan	-11.9	39	Jordan	-6.8
19	Sudan	-11.5	40	Albania	-6.7
20	West Bank & Gaza	-11.4	41	Bahrain	-6.5
21	Tunisia	-11.1	42	Bennin	-6.3

a Estimate.   b 2017

Official reserves[a]

$m, end-2019

1	China	3,222,675		16	France	188,732
2	Japan	1,322,357		17	Mexico	183,043
3	Euro area (19)	913,014		18	Italy	175,122
4	Switzerland	854,812		19	United Kingdom	173,534
5	Russia	554,924		20	Czech Republic	149,855
6	United States	515,785		21	Indonesia	129,178
7	Saudi Arabia	514,926		22	Poland	128,366
8	Taiwan	498,761		23	Israel	126,008
9	India	463,398		24	United Arab Emirates	108,357
10	Hong Kong	441,349		25	Iran[b]	104,555
11	South Korea	408,807		26	Turkey	104,093
12	Brazil	356,879		27	Malaysia	103,625
13	Singapore	285,463		28	Philippines	89,493
14	Thailand	224,338		29	Canada	85,297
15	Germany	223,649		30	Libya	84,648

Official gold reserves

Market prices, $m, end-2019

1	Euro area (19)	526,443		14	Kazakhstan	18,832
2	United States	397,348		15	Portugal	18,688
3	Germany	164,465		16	Uzbekistan	16,411
4	Italy	119,781		17	Saudi Arabia	15,783
5	France	119,007		18	United Kingdom	15,159
6	Russia	110,954		19	Lebanon	14,013
7	China	95,181		20	Spain	13,756
8	Switzerland	50,808		21	Austria	13,679
9	Japan	37,383		22	Poland	11,170
10	India	31,020		23	Belgium	11,109
11	Netherlands	29,920		24	Philippines	9,669
12	Turkey	25,561		25	Algeria	8,483
13	Taiwan	20,635		26	Thailand	7,522

Workers' remittances

Inflows, $m, 2018

1	India	78,790		15	Spain	10,986
2	China	67,414		16	Italy	9,950
3	Mexico	35,768		17	Guatemala	9,438
4	Philippines	33,809		18	Russia	8,610
5	France	27,010		19	Nepal	8,294
6	Egypt	25,516		20	Thailand	7,463
7	Nigeria	24,311		21	South Korea	7,125
8	Pakistan	21,193		22	Poland	7,112
9	Germany	18,032		23	Sri Lanka	7,043
10	Vietnam	16,000		24	Romania	6,984
11	Bangladesh	15,562		25	Lebanon	6,940
12	Ukraine	14,694		26	Morocco	6,918
13	Belgium	12,314		27	Dominican Rep.	6,814
14	Indonesia	11,215		28	United States	6,668

a Foreign exchange, SDRs, IMF position and gold at market prices.   b Estimate.

Exchange rates

The Economist's **Big Mac index**

Local currency under (-)/over (+) valuation against the $ᵃ, %
January 2020

	Big Mac price, $ᵇ

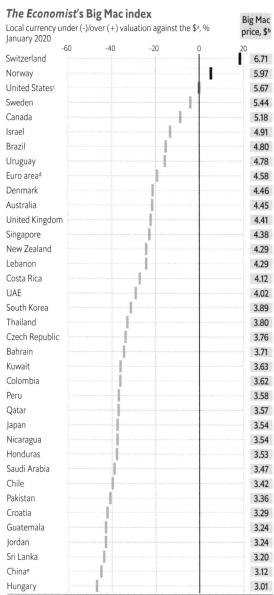

	Big Mac price, $ᵇ
Switzerland	6.71
Norway	5.97
United Statesᶜ	5.67
Sweden	5.44
Canada	5.18
Israel	4.91
Brazil	4.80
Uruguay	4.78
Euro areaᵈ	4.58
Denmark	4.46
Australia	4.45
United Kingdom	4.41
Singapore	4.38
New Zealand	4.29
Lebanon	4.29
Costa Rica	4.12
UAE	4.02
South Korea	3.89
Thailand	3.80
Czech Republic	3.76
Bahrain	3.71
Kuwait	3.63
Colombia	3.62
Peru	3.58
Qatar	3.57
Japan	3.54
Nicaragua	3.54
Honduras	3.53
Saudi Arabia	3.47
Chile	3.42
Pakistan	3.36
Croatia	3.29
Guatemala	3.24
Jordan	3.24
Sri Lanka	3.20
Chinaᵉ	3.12
Hungary	3.01

a Based on purchasing-power parity: local price of a Big Mac burger divided by United States price.   b At market exchange rates.   c Average of four cities.   d Weighted average of prices in euro area.   e Average of five cities.

Inflation

Consumer-price inflation

Highest, 2019 or latest, %

1	Venezuela	19,906.0
2	Zimbabwe	255.3
3	Argentina	53.5
4	South Sudan	51.2
5	Sudan	51.0
6	Iran	41.1
7	Liberia	27.0
8	Haiti	17.3
9	Angola	17.1
10	Ethiopia	15.8
11	Turkey	15.2
12	Sierra Leone	14.8
13	Uzbekistan	14.5
14	Egypt	13.9
15	Nigeria	11.4
16	Yemen	10.0
17	Zambia	9.8
18	Guinea	9.5
19	Malawi	9.4
20	Myanmar	8.6

Lowest, 2019 or latest, %

1	Eritrea	-16.4
2	Burkina Faso	-3.2
3	Niger	-2.5
4	United Arab Emirates	-1.9
5	Saudi Arabia	-1.2
6	Chad	-1.0
7	Benin	-0.9
8	Burundi	-0.7
9	Cayman Islands	-0.6
	Mali	-0.6
	Qatar	-0.6
12	Brunei	-0.5
13	Panama	-0.4
14	Iraq	-0.2
	West Bank & Gaza	-0.2
16	Morocco	0.0
	Somalia	0.0
18	El Salvador	0.1
	Oman	0.1
	Timor-Leste	0.1

Highest average annual consumer-price inflation, 2009–19, %

1	Venezuela	448.7
2	South Sudan	62.5
3	Sudan	31.3
4	Argentina	22.2
5	Iran	21.3
6	Belarus	18.5
7	Yemen	16.9
8	Malawi	16.0
9	Angola	15.8
10	Zimbabwe	15.6
11	Ethiopia	13.5
12	Egypt	12.7
13	Suriname	12.2
14	Uzbekistan	12.0
15	Ukraine	11.9
16	Liberia	11.8
	Nigeria	11.8
18	Guinea	11.7
19	Ghana	11.2
20	Congo-Kinshasa	10.8
21	Turkey	9.8
22	Libya	9.6
	Sierra Leone	9.6
24	Haiti	9.4
25	Zambia	8.9

Lowest average annual consumer-price inflation, 2009–19, %

1	Brunei	-0.1
2	Switzerland	0.0
3	Ireland	0.4
4	Bosnia & Herz.	0.5
	Japan	0.5
6	Burkina Faso	0.6
7	Eritrea	0.7
	Niger	0.7
9	Cyprus	0.8
	Greece	0.8
11	Croatia	1.0
	Israel	1.0
	Senegal	1.0
	Taiwan	1.0
15	Bulgaria	1.1
	Denmark	1.1
	Morocco	1.1
	Puerto Rico	1.1
19	El Salvador	1.2
	Mali	1.2
	Portugal	1.2
	Qatar	1.2
	Slovenia	1.2
	Spain	1.2
	Sweden	1.2
	Togo	1.2

Commodity prices

End 2019, % change on a year earlier		*2014–19, % change*	
1 Palm oil	68.5	**1** Wool (Aus)	47.4
2 Iron ore	33.5	**2** Palm oil	33.1
3 Nickel	32.1	**3** Oil[a]	29.5
4 Beef (Aus)	26.2	**4** Iron ore	28.9
5 Soya oil	25.2	**5** Timber	26.6
6 Oil[a]	24.8	**6** Gold	25.9
7 Timber	21.8	**7** Cotton	13.2
8 Gold	18.7	**8** Silver	7.3
9 Pork	17.1	**9** Tea	6.9
10 Coffee	15.7	**10** Zinc	6.0
11 Silver	15.5	**11** Soya oil	5.5
12 Rice	12.1	**12** Lead	5.1
13 Sugar	11.1	**13** Aluminium	2.5
14 Rubber	11.0	**14** Rice	2.4
Wheat	11.0	**15** Rubber	1.5
16 Soyabeans	6.9	**16** Beef (Aus)	-0.6
17 Corn	3.4	**17** Copper	-1.0
18 Copper	3.2	**18** Corn	-4.3
19 Cocoa	1.6	**19** Wheat	-5.6
20 Beef (US)	-1.0	**20** Nickel	-7.9
21 Soya meal	-2.1	**21** Pork	-9.1
22 Wool (NZ)	-2.5	**22** Sugar	-10.1
23 Aluminium	-3.7	**23** Soyabeans	-10.3
Cotton	-3.7	**24** Cocoa	-18.1
25 Lead	-4.3	**25** Soya meal	-19.1
26 Tea	-4.4	**26** Beef (US)	-20.9

Real[b] residential house prices

Q3 2019[c], % change on a year earlier		*Q3 2014–Q3 2019[c], % change*	
1 Philippines	20.0	**1** Hungary	72.5
2 Portugal	10.5	**2** Philippines	63.1
3 Latvia	10.3	**3** Ireland	42.4
4 Hungary	9.9	**4** Iceland	42.2
5 Greece	9.3	**5** Portugal	41.6
6 Slovakia	8.3	**6** New Zealand	37.2
7 Croatia	7.2	**7** Czech Republic	36.2
8 Luxembourg	7.1	**8** Slovakia	33.8
9 Slovenia	6.5	**9** Chile	33.1
10 Poland	6.2	**10** Hong Kong	30.3
11 Estonia	5.8	Slovenia	30.3
12 Czech Republic	5.7	**12** Luxembourg	29.2
13 Mexico	5.0	**13** Netherlands	28.6
14 Spain	4.4	**14** Canada	27.3
15 Malta	4.2	**15** Germany	26.3
16 Austria	3.8	**16** Bulgaria	25.9
Lithuania	3.8	**17** Malta	25.6
18 Russia	3.6	Spain	25.6
19 Germany	3.4	**19** Sweden	23.3
20 Netherlands	3.3	**20** Austria	23.2

a Brent.   b Deflated by CPI.   c Or latest.

Debt

Highest foreign debt[a]
$bn, 2018

1	China	1,962.3	26	Ukraine	114.5
2	Hong Kong	691.6	27	Romania	112.1
3	Singapore	602.5	28	Vietnam	108.1
4	Brazil	557.8	29	Panama	98.8
5	India	521.4	30	Egypt	98.7
6	Russia	453.9	31	Israel	93.8
7	Mexico	453.0	32	Pakistan	91.0
8	Turkey	445.1	33	Lebanon	79.3
9	South Korea	404.8	34	Philippines	78.8
10	Indonesia	369.8	35	Oman	74.8
11	Argentina	280.5	36	Iraq	71.4
12	United Arab Emirates	238.1	37	Peru	66.7
13	Poland	229.0	38	Sudan	59.1
14	Malaysia	223.5	39	Kuwait	57.7
15	Saudi Arabia	218.9	40	Bahrain	56.2
16	Czech Republic	194.1	41	Angola	54.6
17	Taiwan	191.2	42	Sri Lanka	52.6
18	Qatar	187.4	43	Bangladesh	52.1
19	Chile	185.5	44	Morocco	49.0
20	South Africa	179.3	45	Nigeria	47.0
21	Thailand	169.2	46	Ecuador	45.0
22	Kazakhstan	156.9	47	Croatia	44.0
23	Venezuela	154.9	48	Uruguay	42.2
24	Hungary	152.7	49	Bulgaria	39.9
25	Colombia	134.9	50	Belarus	38.8

Highest foreign debt burden[a]
Total foreign debt as % of GDP, 2018

1	Mongolia	224.8	21	Ukraine	87.5
2	Sudan	191.4	22	Tunisia	87.2
3	Hong Kong	190.7	23	Laos	86.8
4	Singapore	161.4	24	Tajikistan	79.4
5	Panama	151.9	25	Czech Republic	79.2
6	Bahrain	149.1	26	Mauritius	79.0
7	Montenegro	144.4	27	Bosnia & Herz.	76.1
8	Lebanon	140.1	28	Jordan	75.9
9	Jamaica	106.9	29	Croatia	72.1
10	Mozambique	103.4	30	Zambia	71.6
11	Kyrgyzstan	100.3	31	Uruguay	70.8
12	Qatar	97.9	32	Macedonia	69.2
13	Georgia	97.3	33	Serbia	68.7
14	Mauritania	97.2	34	El Salvador	67.2
15	Hungary	96.9	35	Albania	67.0
16	Oman	94.4	36	Belarus	65.1
17	Papua New Guinea	90.5	37	Venezuela	63.9
18	Kazakhstan	89.8	38	Moldova	63.8
19	Armenia	88.6	39	Cambodia	62.4
	Nicaragua	88.6	40	Malaysia	62.3

a Foreign debt is debt owed to non-residents and repayable in foreign currency; the figures shown include liabilities of government, public and private sectors. Longer-established developed countries have been excluded.

Highest foreign debt[a]

As % of exports of goods and services, 2018

1	Sudan	1,064.7		15	Kenya	218.0
2	Venezuela	433.6		16	Sierra Leone	209.1
3	Mongolia	350.6		17	Tanzania	207.0
4	Ethiopia	341.7		18	Colombia	202.8
5	Argentina	335.9		19	Central African Rep.	201.9
6	Panama	315.3		20	Chile	199.1
7	Lebanon	271.3		21	Malawi	192.3
8	Montenegro	241.4		22	Brazil	191.6
9	Laos	236.5		23	Jamaica	191.2
10	Mozambique	234.8		24	Sri Lanka	190.9
11	Rwanda	232.7		25	Turkey	189.7
12	Uruguay	231.8		26	Yemen	189.1
13	Mauritania	227.4		27	Zambia	189.0
14	Kazakhstan	224.2		28	Zimbabwe	184.4

Highest debt service ratio[b]

Average, %, 2018

1	Mongolia	96.3		15	Croatia	31.2
2	Syria	67.0		16	Papua New Guinea	28.8
3	Montenegro	52.3		17	Namibia	28.1
4	Lebanon	49.0		18	Venezuela	27.9
5	Kazakhstan	48.0		19	El Salvador	27.0
6	Argentina	44.8		20	Sri Lanka	26.8
7	Chile	44.6		21	Panama	26.4
8	Sudan	39.5		22	Ethiopia	24.5
9	Colombia	36.8		23	Armenia	23.7
10	Hungary	36.0		24	Jamaica	23.1
11	Turkey	35.9		25	Mauritius	22.9
12	Bahrain	35.5		26	Georgia	22.8
13	Brazil	32.8		27	Angola	21.9
14	Ecuador	32.7		28	South Africa	21.4

Household debt[c]

As % of net disposable income, 2018

1	Denmark	281.3		14	New Zealand[d]	126.6
2	Netherlands	239.5		15	France	120.7
3	Norway	239.2		16	Belgium	115.0
4	Switzerland	223.0		17	Japan[d]	108.0
5	Australia	216.8		18	Spain	107.0
6	Sweden	188.9		19	Greece	105.6
7	Luxembourg	186.4		20	United States	105.4
8	South Korea	184.2		21	Germany	95.3
9	Canada	181.8		22	Austria	90.3
10	Finland	144.9		23	Italy	86.8
11	United Kingdom	141.2		24	Slovakia	79.4
12	Ireland	140.4		25	Estonia	79.0
13	Portugal	127.4		26	Chile[d]	70.2

b Debt service is the sum of interest and principal repayments (amortisation) due on outstanding foreign debt. The debt service ratio is debt service as a percentage of exports of goods, non-factor services, primary income and workers' remittances.
c OECD countries. d 2017

Aid

Largest recipients of bilateral and multilateral aid
$bn, 2018

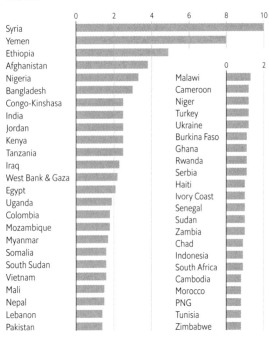

Syria
Yemen
Ethiopia
Afghanistan
Nigeria
Bangladesh
Congo-Kinshasa
India
Jordan
Kenya
Tanzania
Iraq
West Bank & Gaza
Egypt
Uganda
Colombia
Mozambique
Myanmar
Somalia
South Sudan
Vietnam
Mali
Nepal
Lebanon
Pakistan

Malawi
Cameroon
Niger
Turkey
Ukraine
Burkina Faso
Ghana
Rwanda
Serbia
Haiti
Ivory Coast
Senegal
Sudan
Zambia
Chad
Indonesia
South Africa
Cambodia
Morocco
PNG
Tunisia
Zimbabwe

$ per person, 2018

1	Syria	591.0	21	Mongolia	104.6
2	West Bank & Gaza	490.3	22	Gambia, The	102.0
3	Yemen	280.2	23	Afghanistan	101.9
4	Jordan	253.5	24	Mauritania	101.6
5	Montenegro	248.8	25	Papua New Guinea	91.3
6	Maldives	231.0	26	Rwanda	91.0
7	Lebanon	207.3	27	Haiti	89.2
8	Kosovo	186.9	28	Macedonia	81.6
9	Timor-Leste	159.8	29	Guinea-Bissau	81.3
10	Georgia	157.9	30	Laos	80.0
11	Serbia	152.9	31	Mali	78.2
12	South Sudan	143.7	32	Lesotho	72.0
13	Central African Rep.	140.5	33	Malawi	70.1
14	Guyana	131.5	34	Tunisia	69.6
15	Fiji	125.8	35	Honduras	68.8
16	Albania	119.6	36	Sierra Leone	66.1
17	Liberia	118.4	37	Cuba	65.9
18	Bosnia & Herz.	105.8	38	Kyrgyzstan	65.5
19	Eswatini	105.2	39	Moldova	65.1
20	Somalia	104.8	40	Bolivia	64.0

Largest bilateral and multilateral donors[a]

2018

		$bn	As % of GNI			$bn	As % of GNI
1	United States	33.8	0.2		Switzerland	3.1	0.5
2	Germany	25.7	0.7	16	Denmark	2.6	0.7
3	United Kingdom	19.5	0.7	17	Spain	2.5	0.2
4	France	12.8	0.4	18	South Korea	2.4	0.1
5	Japan	10.1	0.2	19	Belgium	2.3	0.4
6	Turkey	8.6	1.1	20	Austria	1.2	0.3
7	Sweden	5.8	1.0	21	Finland	1.0	0.4
8	Netherlands	5.6	0.6		Russia	1.0	0.1
9	Italy	5.1	0.3	23	Ireland	0.9	0.3
10	Saudi Arabia	4.8	0.6	24	Poland	0.8	0.1
11	Canada	4.6	0.3	25	New Zealand	0.6	0.2
12	Norway	4.3	1.0	26	Luxembourg	0.5	1.0
13	United Arab			27	Israel	0.4	0.1
	Emirates	4.1	1.0		Portugal	0.4	0.2
14	Australia	3.1	0.2				

Biggest changes to aid

2018 compared with 2014, $m

Increases			Decreases		
1	Yemen	6,821.8	1	Vietnam	-2,582.3
2	Syria	5,816.5	2	Turkey	-2,259.5
3	Ethiopia	1,346.1	3	Pakistan	-2,253.8
4	Indonesia	1,330.5	4	Egypt	-1,473.9
5	Iraq	930.6	5	Morocco	-1,428.5
6	Nigeria	822.9	6	Afghanistan	-1,154.1
7	Serbia	700.6	7	Thailand	-777.7
8	Bangladesh	617.6	8	Sri Lanka	-749.6
9	Lebanon	598.7	9	India	-537.8
10	Nepal	566.9	10	Brazil	-485.0
11	Colombia	538.5	11	Sierra Leone	-408.1
12	Cuba	485.3	12	South Sudan	-386.8
13	Chad	483.5	13	Moldova	-286.9
14	Somalia	463.8	14	Mozambique	-286.2
15	Malawi	339.8	15	Bosnia & Herz.	-278.6
16	Cameroon	308.0	16	Mexico	-275.1
17	Uganda	307.1	17	West Bank & Gaza	-247.4
18	Myanmar	303.2	18	Kosovo	-234.5
19	Niger	278.5	19	Kyrgyzstan	-213.0
20	Mali	256.4	20	Tanzania	-197.3
21	Panama	231.5	21	Ukraine	-182.7
22	Ecuador	231.3	22	Liberia	-178.8
23	Uzbekistan	231.2	23	Jordan	-173.2
24	China	210.0	24	Kenya	-172.6
25	Papua New Guinea	203.6	25	South Africa	-162.6
26	Mauritania	186.5	26	Philippines	-136.9
27	El Salvador	150.5	27	Azerbaijan	-132.8
28	Gambia, The	132.3	28	Armenia	-128.3
29	Guatemala	113.8	29	Tunisia	-117.4

a Countries that report figures to the OECD.

Industry and services

Largest industrial output
$bn, 2018

1	China	5,532	23	Netherlands	164
2	United States[a]	3,548	24	Iran	158
3	Japan[a]	1,416	25	Ireland	141
4	Germany	1,084	26	Norway	139
5	India	727	27	Malaysia	137
6	South Korea	569	28	Iraq	126
7	Russia	532	29	Sweden	125
8	United Kingdom	500	30	Argentina	119
9	France	469	31	Austria	117
10	Italy	446		Qatar	117
11	Indonesia	414	33	Belgium	104
12	Saudi Arabia	390	34	Nigeria	102
13	Canada[b]	385		Philippines	102
14	Mexico	378	36	South Africa	95
15	Australia	346	37	Singapore	92
16	Brazil	345	38	Chile	89
17	Spain	284	39	Colombia	88
18	Turkey	227		Egypt	88
19	United Arab Emirates	194	41	Kuwait	84
20	Thailand	177		Vietnam	84
21	Switzerland	176	43	Czech Republic	79
22	Poland	168	44	Bangladesh	78

Highest growth in industrial output
Average annual % increase in real terms, 2008–18

1	Ethiopia	23.6	11	Mongolia	11.3
2	Zimbabwe	22.3	12	Maldives	10.2
3	Laos	14.5	13	China	9.9
4	Panama	14.3	14	West Bank & Gaza[c]	9.5
5	Ghana	14.1	15	Guyana	8.8
6	Bangladesh	13.2	16	Vietnam	8.6
7	Cambodia	13.1	17	Moldova	8.4
8	Liberia	12.8	18	Burkina Faso	8.2
9	Myanmar	12.3	19	Kenya	8.0
10	Congo-Kinshasa	12.2	20	Bolivia	7.9

Lowest growth in industrial output
Average annual % change in real terms, 2008–18

1	Sudan	-25.0	12	Tunisia	-4.3
2	Timor-Leste[d]	-11.3	13	Congo-Brazzaville	-4.2
3	South Sudan[e]	-10.5	14	Spain	-4.0
4	Andorra[d]	-7.8	15	Algeria	-3.7
5	Equatorial Guinea	-7.5	16	Croatia	-2.9
6	Trinidad & Tobago	-5.6		Norway	-2.9
7	Bermuda[f]	-5.5	18	Isle of Man[d]	-2.8
8	Ukraine	-5.3	19	Azerbaijan	-2.6
9	Greece	-5.0	20	Canada[e]	-2.5
10	Cyprus	-4.8	21	Italy	-2.4
11	Yemen	-4.4			

a 2017   b 2015   c 2008–16   d 2008–17   e 2008–15   f 2008–12

Largest manufacturing output
$bn, 2018

1	China	4,003		21	Netherlands	101
2	United States[a]	2,173			Saudi Arabia	101
3	Japan[a]	1,007		23	Poland	98
4	Germany	805		24	Australia	83
5	South Korea	441		25	Austria	77
6	India	403			Malaysia	77
7	Italy	314		27	Singapore	76
8	France	270		28	Sweden	74
9	United Kingdom	253		29	Belgium	67
10	Mexico	211		30	Argentina	66
11	Indonesia	207		31	Philippines	63
12	Russia	204		32	Czech Republic	57
13	Taiwan	198		33	Iran[a]	54
14	Brazil	181		34	Bangladesh	49
15	Canada[b]	161		35	Puerto Rico	48
16	Spain	159			Romania	48
17	Turkey	147		37	Algeria	47
18	Thailand	136		38	Denmark	46
19	Switzerland	128		39	Israel	44
20	Ireland	124		40	South Africa	43

Largest services output
$bn, 2017

1	United States[c]	14,300		27	Thailand	256
2	China	6,320		28	Norway	230
3	Japan[c]	3,400		29	Singapore	228
4	Germany	2,270		30	Venezuela[d]	226
5	United Kingdom	1,860		31	Israel[b]	222
6	France	1,810		32	Denmark	216
7	Brazil	1,300		33	South Africa	215
8	Italy	1,290		34	Nigeria	210
9	India	1,270		35	Philippines	188
10	Canada[b]	1,160		36	Ireland	187
11	Australia	886		37	United Arab Emirates	179
	Russia	886		38	Colombia	172
13	Spain	869		39	Pakistan	162
14	South Korea	809		40	Chile	160
15	Mexico	701			Malaysia	160
16	Netherlands	584		42	Finland	150
17	Switzerland	485		43	Portugal	143
18	Turkey	454		44	Greece	139
19	Indonesia	443		45	Bangladesh	134
20	Argentina	363		46	Romania	119
21	Saudi Arabia	359		47	Czech Republic	118
22	Sweden	349		48	New Zealand[b]	117
23	Belgium	341		49	Kazakhstan	92
24	Poland	306		50	Vietnam[c]	84
25	Iran	287		51	Hungary	77
26	Austria	261		52	Algeria	75

a 2017   b 2015   c 2016   d 2014

Agriculture and fisheries

Largest agricultural output
$bn, 2018

1	China	978.5	16	Spain	39.7
2	India	397.0	17	Vietnam	36.0
3	United States[a]	178.6	18	Bangladesh	35.8
4	Indonesia	133.5	19	Australia	35.2
5	Nigeria	84.2	20	South Korea	32.1
6	Brazil	81.5	21	Argentina	31.7
7	Pakistan	71.9	22	Philippines	30.7
8	Japan[a]	57.8	23	Germany	30.3
9	Russia	52.2	24	Kenya	30.1
10	France	45.1	25	Egypt	28.2
11	Turkey	44.9	26	Malaysia	27.0
12	Iran[a]	43.1	27	Canada[b]	26.6
13	Mexico	41.3	28	Ethiopia	26.3
14	Thailand	41.0	29	Venezuela[c]	24.2
15	Italy	40.4			

Most economically dependent on agriculture
% of GDP from agriculture, 2018

1	Sierra Leone	58.9	15	Burkina Faso	28.0
2	Guinea-Bissau	47.5	16	Malawi[a]	26.1
3	Chad	44.9	17	Mauritania	25.9
4	Niger	39.2	18	Nepal	25.3
5	Mali	38.7	19	Myanmar	24.6
6	Liberia	37.3	20	Mozambique	24.5
7	Kenya	34.2	21	Guinea	24.3
8	Sudan	31.5	22	Uganda	24.2
9	Central African Rep.	31.2	23	Madagascar	23.8
	Ethiopia	31.2	24	Togo	23.4
11	Burundi	29.0	25	North Korea	23.3
	Rwanda	29.0	26	Pakistan	22.9
13	Uzbekistan	28.8	27	Benin	22.6
14	Tanzania[a]	28.7	28	Cambodia	22.0

Least economically dependent on agriculture
% of GDP from agriculture, 2018

1	Macau	0.0	14	Switzerland	0.7
	Singapore	0.0		United Arab Emirates	0.7
3	Hong Kong	0.1	16	Germany	0.8
4	Luxembourg	0.2		Puerto Rico	0.8
	Qatar	0.2	18	Bahamas	0.9
6	Bahrain	0.3		Ireland	0.9
7	Cayman Islands[a]	0.4		Malta	0.9
	Curaçao	0.4		United States[a]	0.9
	Isle of Man[a]	0.4	22	Brunei	1.0
	Kuwait	0.4		Denmark	1.0
11	Andorra[a]	0.5		Trinidad & Tobago	1.0
	Belgium	0.5	25	Austria	1.1
13	United Kingdom	0.6		Israel	1.1

a 2017   b 2015   c 2014

Fisheries and aquaculture production
Fish, crustaceans and molluscs, million tonnes, 2018

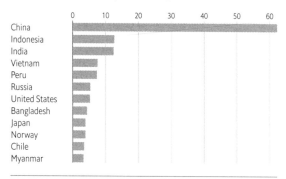

Biggest producers
'000 tonnes, 2017

Cereals

1	China	619,879	6	Indonesia	110,073
2	United States	466,847	7	Argentina	76,397
3	India	313,640	8	France	68,730
4	Russia	131,295	9	Ukraine	60,686
5	Brazil	117,979	10	Bangladesh	58,496

Meat

1	China	88,420	6	India	7,366
2	United States	45,799	7	Mexico	6,808
3	Brazil	27,694	8	Spain	6,662
4	Russia	10,319	9	Argentina	5,699
5	Germany	8,281	10	France	5,563

Fruit

1	China	226,083	6	Turkey	21,341
2	India	92,909	7	Indonesia	19,416
3	Brazil	39,168	8	Spain	17,829
4	United States	25,620	9	Philippines	16,703
5	Mexico	21,542	10	Iran	16,515

Vegetables

1	China	555,572	6	Egypt	16,154
2	India	129,470	7	Mexico	16,117
3	United States	32,937	8	Iran	16,082
4	Turkey	26,791	9	Vietnam	15,760
5	Nigeria	16,406	10	Spain	13,802

Roots and tubers

1	China	146,726	5	Thailand	31,365
2	Nigeria	115,835	6	Ghana	28,401
3	India	54,236	7	Brazil	23,188
4	Congo-Kinshasa	32,721	8	Indonesia	22,553

Commodities

Wheat

Top 10 producers, 2018–19
'000 tonnes

1	EU28	136,863
2	China	131,430
3	India	99,870
4	Russia	71,685
5	United States	51,306
6	Canada	32,201
7	Pakistan	25,100
8	Ukraine	25,057
9	Argentina	19,500
10	Turkey	19,000

Top 10 consumers, 2018–19
'000 tonnes

1	China	128,000
2	EU28	126,500
3	India	98,000
4	Russia	39,500
5	United States	30,024
6	Pakistan	25,400
7	Egypt	20,400
8	Turkey	19,700
9	Iran	16,400
10	Brazil	12,100

Rice[a]

Top 10 producers, 2018–19
'000 tonnes

1	China	148,490
2	India	116,480
3	Indonesia	36,700
4	Bangladesh	34,909
5	Vietnam	27,344
6	Thailand	20,340
7	Myanmar	13,200
8	Philippines	11,732
9	Japan	7,657
10	Pakistan	7,300

Top 10 consumers, 2018–19
'000 tonnes

1	China	142,720
2	India	99,160
3	Indonesia	38,100
4	Bangladesh	35,400
5	Vietnam	21,200
6	Philippines	14,100
7	Thailand	11,500
8	Myanmar	10,250
9	Japan	8,400
10	Brazil	7,450

Sugar[b]

Top 10 producers, 2018
'000 tonnes

1	India	33,300
2	Brazil	29,290
3	EU28	18,180
4	Thailand	15,440
5	China	10,710
6	United States	7,830
7	Pakistan	6,280
8	Russia	6,180
9	Mexico	5,920
10	Australia	4,640

Top 10 consumers, 2018
'000 tonnes

1	India	25,390
2	EU28	17,940
3	China	16,100
4	Brazil	10,470
5	United States	10,190
6	Indonesia	6,890
7	Russia	5,820
8	Pakistan	5,250
9	Mexico	4,270
10	Egypt	3,300

Coarse grains[c]

Top 5 producers, 2018–19
'000 tonnes

1	United States	377,905
2	China	264,512
3	EU28	148,107
4	Brazil	104,334
5	Argentina	59,223

Top 5 consumers, 2018–19
'000 tonnes

1	United States	326,242
2	China	287,370
3	EU28	168,570
4	Brazil	70,983
5	Mexico	50,505

a Milled. b Raw. c Includes: maize (corn), barley, sorghum, oats, rye, millet, triticale and other. d Tonnes at 65 degrees brix.

Tea

Top 10 producers, 2018
'000 tonnes

1	China	2,610
2	India	1,345
3	Kenya	493
4	Sri Lanka	304
5	Turkey	270
	Vietnam	270
7	Indonesia	141
8	Iran	109
	Myanmar	109
10	Japan	83

Top 10 consumers, 2018
'000 tonnes

1	China	2,285
2	India	1,103
3	Turkey	323
4	EU28	223
5	Pakistan	192
6	Russia	150
7	United States	116
8	Japan	105
9	Egypt	95
10	Bangladesh	90

Coffee

Top 10 producers, 2018
'000 tonnes

1	Brazil	3,777
2	Vietnam	1,792
3	Colombia	832
4	Indonesia	589
5	Ethiopia	467
6	Honduras	440
7	India	360
8	Uganda	282
9	Mexico	261
10	Peru	248

Top 10 consumers, 2018–19
'000 tonnes

1	EU28	2,740
2	United States	1,676
3	Brazil	1,338
4	Japan	454
5	Indonesia	294
6	Russia	278
7	Canada	237
8	Ethiopia	230
9	Philippines	201
10	Vietnam	174

Cocoa

Top 10 producers, 2018–19
'000 tonnes

1	Ivory Coast	2,154
2	Ghana	812
3	Ecuador	322
4	Cameroon	280
5	Nigeria	250
6	Indonesia	200
7	Brazil	176
8	Peru	130
9	Dominican Rep.	75
10	Colombia	60

Top 10 consumers, 2018–19
'000 tonnes

1	EU28	1,585
2	Ivory Coast	605
3	Netherlands	600
4	Indonesia	487
5	Germany	445
6	United States	400
7	Malaysia	327
8	Ghana	320
9	Brazil	235
10	France	150

Orange juice[d]

Top 5 producers, 2018–19
'000 tonnes

1	Brazil	1,327
2	United States	329
3	Mexico	195
4	EU28	97
5	China	45

Top 5 consumers, 2018–19
'000 tonnes

1	EU28	695
2	United States	530
3	China	112
4	Canada	87
5	Japan	70

Copper

Top 10 producers[a], 2018		*Top 10 consumers[b], 2018*	
'000 tonnes		*'000 tonnes*	
1 Chile	5,832	1 China	12,482
2 Peru	2,437	2 EU28	3,317
3 China	1,507	3 United States	1,814
4 Congo-Kinshasa	1,225	4 Germany	1,200
5 United States	1,216	5 Japan	1,039
6 EU28	924	6 South Korea	717
7 Australia	913	7 Italy	552
8 Zambia	857	8 India	512
9 Russia	773	9 Turkey	482
10 Mexico	751	10 Mexico	407

Lead

Top 10 producers[a], 2018		*Top 10 consumers[b], 2018*	
'000 tonnes		*'000 tonnes*	
1 China	2,214	1 China	5,235
2 Australia	469	2 EU28	1,782
3 Peru	289	3 United States	1,613
4 United States	260	4 South Korea	615
5 Mexico	235	5 India	569
6 Russia	215	6 Germany	389
7 EU28	186	7 Japan	271
8 India	185	8 Spain	257
9 Bolivia	112	9 Brazil	248
10 Kazakhstan	86	10 United Kingdom	236

Zinc

Top 10 producers[a], 2018		*Top 10 consumers[c], 2018*	
'000 tonnes		*'000 tonnes*	
1 China	4,193	1 China	6,179
2 Peru	1,475	2 EU28	1,982
3 Australia	1,112	3 United States	867
4 United States	838	4 South Korea	716
5 India	765	5 India	714
6 EU28	708	6 Japan	482
7 Mexico	637	7 Germany	444
8 Bolivia	520	8 Spain	287
9 Kazakhstan	345	9 Italy	280
10 Russia	315	10 Turkey	248

Tin

Top 5 producers[a], 2018		*Top 5 consumers[b], 2018*	
'000 tonnes		*'000 tonnes*	
1 China	127.0	1 China	174.2
2 Indonesia	84.0	2 EU28	59.5
3 Myanmar	45.9	3 United States	34.7
4 Peru	18.6	4 Japan	28.1
5 Brazil	18.0	5 Germany	20.2

Nickel

Top 10 producers[a], 2018
'000 tonnes

1	Indonesia	647.7
2	Philippines	390.0
3	Russia	218.3
4	New Caledonia	216.2
5	Canada	180.0
6	Australia	160.9
7	China	99.0
8	Brazil	65.2
9	EU28	62.2
10	Cuba	52.1

Top 10 consumers[b], 2018
'000 tonnes

1	China	1,096.4
2	EU28	332.7
3	Japan	175.2
4	Indonesia	173.0
5	United States	136.4
6	South Korea	117.5
7	Taiwan	87.7
8	India	72.1
9	Germany	60.6
10	Italy	58.5

Aluminium

Top 10 producers[d], 2018
'000 tonnes

1	China	35,802
2	Russia	3,621
3	India	2,934
4	Canada	2,923
5	United Arab Emirates	2,635
6	EU28	2,321
7	Australia	1,574
8	Vietnam	1,310
9	Norway	1,295
10	Bahrain	1,011

Top 10 consumers[e], 2018
'000 tonnes

1	China	35,521
2	EU28	7,535
3	United States	4,630
4	Germany	2,139
5	Japan	1,979
6	India	1,750
7	Vietnam	1,253
8	South Korea	1,151
9	Turkey	954
10	Italy	951

Precious metals

Gold[a]
Top 10 producers, 2018
tonnes

1	China	401.1
2	Australia	313.0
3	Russia	279.9
4	United States	225.6
5	Canada	186.9
6	Ghana	159.0
7	Peru	142.6
8	Mexico	118.4
9	South Africa	117.1
10	Indonesia	111.9

Silver[a]
Top 10 producers, 2018
tonnes

1	Mexico	5,624
2	Peru	4,161
3	China	3,574
4	EU28	2,076
5	Poland	1,409
6	Russia	1,350
7	Chile	1,243
8	Australia	1,220
9	Bolivia	1,191
10	Kazakhstan	969

Platinum
Top 3 producers, 2018
tonnes

1	South Africa	137.0
2	Russia	22.0
3	Zimbabwe	15.0

Palladium
Top 3 producers, 2018
tonnes

1	Russia	90.0
2	South Africa	80.6
3	Canada	20.0

a Mine production. b Refined consumption. c Slab consumption.
d Primary refined production. e Primary refined consumption.

Rubber (natural and synthetic)

Top 10 producers, 2018		*Top 10 consumers, 2018*	
'000 tonnes		*'000 tonnes*	
1 Thailand	5,417	1 China	9,909
2 China	3,944	2 EU28	3,803
3 Indonesia	3,486	3 United States	2,870
4 EU28	2,456	4 India	1,911
5 United States	2,284	5 Japan	1,594
6 South Korea	1,592	6 Thailand	1,336
7 Russia	1,575	7 Malaysia	1,050
8 Japan	1,569	8 Indonesia	1,041
9 Vietnam	1,142	9 Brazil	853
10 India	1,036	10 Russia	645

Cotton

Top 10 producers, 2018–19		*Top 10 consumers, 2018–19*	
'000 tonnes		*'000 tonnes*	
1 China	6,040	1 China	8,250
2 India	5,350	2 India	5,400
3 United States	3,999	3 Pakistan	2,358
4 Brazil	2,726	4 Bangladesh	1,579
5 Pakistan	1,670	5 Turkey	1,555
6 Turkey	977	6 Vietnam	1,506
7 Uzbekistan	641	7 Brazil	730
8 Australia	485	8 Indonesia	700
9 Mexico	414	9 United States	644
10 Greece	308	10 Uzbekistan	630

Major oil seeds[a]

Top 5 producers, 2018–19		*Top 5 consumers, 2018–19*	
'000 tonnes		*'000 tonnes*	
1 Brazil	129,886	1 China	151,124
2 United States	107,423	2 United States	71,066
3 China	62,629	3 Argentina	55,959
4 Argentina	57,859	4 EU28	51,781
5 India	35,594	5 Brazil	50,978

Major vegetable oils[b]

Top 5 producers, 2018–19		*Top 5 consumers, 2018–19*	
'000 tonnes		*'000 tonnes*	
1 Indonesia	42,500	1 China	33,799
2 China	22,251	2 EU28	24,388
3 Malaysia	19,898	3 India	20,560
4 EU28	15,939	4 United States	14,705
5 United States	11,933	5 Indonesia	14,318

a Soyabeans, rapeseed (canola), cottonseed, sunflowerseed and groundnuts (peanuts).   b Palm, soyabean, rapeseed and sunflowerseed oil.
c Includes crude oil, shale oil, oil sands and natural gas liquids.   d Opec member.
e Opec membership suspended 30 November 2016.   f Left Opec 1 January 2019.

Oil[c]

Top 10 producers, 2019		*Top 10 consumers, 2019*	
'000 barrels per day		*'000 barrels per day*	
1 United States	17,045	1 United States	19,400
2 Saudi Arabia[d]	11,832	2 China	14,056
3 Russia	11,540	3 India	5,271
4 Canada	5,651	4 Japan	3,812
5 Iraq[d]	4,779	5 Saudi Arabia[d]	3,788
6 United Arab Emirates[d]	3,998	6 Russia	3,317
7 China	3,836	7 South Korea	2,760
8 Iran[d]	3,535	8 Canada	2,403
9 Kuwait[d]	2,996	9 Brazil	2,398
10 Brazil	2,877	10 Germany	2,281

Natural gas

Top 10 producers, 2019		*Top 10 consumers, 2019*	
Billion cubic metres		*Billion cubic metres*	
1 United States	920.9	1 United States	846.6
2 Russia	679.0	2 Russia	444.3
3 Iran[d]	244.2	3 China	307.3
4 Qatar[f]	178.1	4 Iran[d]	223.6
5 China	177.6	5 Canada	120.3
6 Canada	173.1	6 Saudi Arabia[d]	113.6
7 Australia	153.5	7 Japan	108.1
8 Norway	114.4	8 Mexico	90.7
9 Saudi Arabia[d]	113.6	9 Germany	88.7
10 Algeria[d]	86.2	10 United Kingdom	78.8

Coal

Top 10 producers, 2019		*Top 10 consumers, 2019*	
Exajoules		*Exajoules*	
1 China	79.82	1 China	81.67
2 Indonesia[e]	15.05	2 India	18.62
3 United States	14.30	3 United States	11.34
4 Australia	13.15	4 Japan	4.91
5 India	12.73	5 South Africa	3.81
6 Russia	9.20	6 Russia	3.63
7 South Africa	6.02	7 South Korea	3.44
8 Colombia	2.37	8 Indonesia[e]	3.41
9 Kazakhstan	2.08	9 Germany	2.30
10 Poland	1.87	10 Vietnam	2.07

Oil reserves[c]

Top proved reserves, end 2019			
% of world total			
1 Venezuela[d]	17.5	6 Russia	6.2
2 Saudi Arabia[d]	17.2	7 Kuwait[d]	5.9
3 Canada	9.8	8 United Arab Emirates[d]	5.6
4 Iran[d]	9.0	9 United States	4.0
5 Iraq[d]	8.4	10 Libya[d]	2.8

Energy

Largest producers
Million tonnes of oil equivalent, 2017

1	China	2,825	16	Algeria	171
2	United States	2,223	17	Kuwait	169
3	Russia	1,553	18	Mexico	168
4	Saudi Arabia	727	19	Venezuela	157
5	Canada	552	20	Nigeria	150
6	Iran	459		South Africa	150
7	India	446	22	United Kingdom	133
8	Australia	417	23	Colombia	129
9	Indonesia	359	24	France	123
10	Brazil	283	25	Germany	120
11	Qatar	256	26	Malaysia	109
12	United Arab Emirates	249	27	Angola	96
13	Norway	248	28	Egypt	87
14	Iraq	241		Turkmenistan	87
15	Kazakhstan	203	30	Oman	81

Largest consumers
Million tonnes of oil equivalent, 2017

1	China	3,517	16	Italy	171
2	United States	2,464	17	Turkey	162
3	Russia	828	18	Australia	153
4	India	769	19	Spain	145
5	Japan	494	20	South Africa	143
6	Canada	380	21	Thailand	139
7	Germany	353	22	Taiwan	118
8	Brazil	318		United Arab Emirates	118
9	South Korea	312	24	Poland	110
10	Iran	293	25	Egypt	101
11	Saudi Arabia	277	26	Argentina	98
12	France	260		Netherlands	98
13	United Kingdom	207	28	Kazakhstan	92
14	Mexico	199		Singapore	92
15	Indonesia	181		Ukraine	92

Energy efficiency[a]
GDP per unit of energy use, 2017

Most efficient			Least efficient		
1	Chad	245.7	1	Trinidad & Tobago	1.8
2	Cyprus	123.7	2	Turkmenistan	2.0
3	Rwanda	59.6	3	Czech Republic	2.1
4	Burundi	58.3		Virgin Islands (US)	2.1
5	Macau	56.8	5	Libya	2.6
6	Uganda	35.3	6	Cuba	3.2
7	Eritrea	34.6		Iceland	3.2
8	Sierra Leone	32.2	8	Kyrgyzstan	3.7
9	Timor-Leste	30.4	9	Bahrain	3.8
10	Guinea-Bissau	29.0	10	Ukraine	3.9
11	Gambia, The	28.2	11	Canada	4.5

a 2015 PPP $ per kg of oil equivalent.   b Coal, gas and oil.

Net energy importers
% of commercial energy use, 2017

Highest		Lowest	
1 Bermuda	100.0	1 Chad	-6,405.4
Liberia	100.0	2 Timor-Leste	-3,312.3
Macau	100.0	3 South Sudan	-1,619.2
West Bank & Gaza	100.0	4 Equatorial Guinea	-951.1
5 Benin	99.9	5 Angola	-945.4
Hong Kong	99.9	6 Gabon	-553.6
Maldives	99.9	7 Congo-Brazzaville	-521.0
8 Guinea-Bissau	99.8	8 Papua New Guinea	-461.0
9 Eritrea	99.6	9 Norway	-419.0
Gambia, The	99.6	10 Iraq	-411.7
Singapore	99.6	11 Mongolia	-390.9
Virgin Islands (US)	99.6	12 Qatar	-369.2

Largest consumption per person
Kg of oil equivalent, 2017

1 Qatar	20,638.2	12 Virgin Islands (US)	8,166.2
2 Iceland	16,705.6	13 Luxembourg	7,934.1
3 Trinidad & Tobago	16,649.0	14 Turkmenistan	7,606.0
4 Singapore	16,034.2	15 United States	7,593.8
5 United Arab Emirates	12,520.8	16 Malta	7,521.6
6 Bahrain	12,149.2	17 New Caledonia	7,130.4
7 Brunei	10,405.5	18 Australia	6,274.9
8 Canada	10,366.1	19 Oman	6,197.6
9 Kuwait	9,701.1	20 South Korea	6,113.6
10 Norway	8,994.3	21 Belgium	5,864.2
11 Saudi Arabia	8,419.8	22 Netherlands	5,767.8

Sources of electricity
% of total, 2017

Fossil fuels[b]		Nuclear power	
1 Bermuda	100.00	1 France	71.5
Liberia	100.00	2 Ukraine	55.2
Macau	100.00	3 Slovakia	54.1
Oman	100.00	4 Belgium	49.7
5 Turkmenistan	99.99	5 Hungary	48.8

Hydropower		Renewables excl. hydropower	
1 Albania	100.0	1 Denmark	74.1
Lesotho	100.0	2 Lithuania	65.3
3 Paraguay	99.8	3 Uruguay	43.5
4 Congo-Kinshasa	99.6	4 Portugal	30.6
5 Central African Rep.	99.1	5 Ireland	28.6

Wind power		Biomass and waste	
1 Denmark	48.4	1 Eswatini	35.5
2 Lithuania	44.0	2 Denmark	23.2
3 Uruguay	27.9	3 Guatemala	20.0
4 Ireland	25.5	4 Lithuania	19.1
5 Portugal	22.1	5 Finland	18.9

Labour markets

Labour-force participation
% of working-age population[a] working or looking for work, 2019 or latest

Highest			Lowest		
1	Qatar	86.8	1	Yemen	38.0
2	Madagascar	86.1	2	Jordan	39.3
3	Nepal	83.8		Puerto Rico	39.3
4	Rwanda	83.7	4	Algeria	41.2
5	Tanzania	83.4	5	Tajikistan	42.0
6	Zimbabwe	83.1	6	Iraq	43.0
7	Cambodia	82.3	7	Moldova	43.1
8	United Arab Emirates	82.1	8	Syria	44.1
9	North Korea	80.4	9	Iran	44.7
10	Ethiopia	79.6	10	Morocco	45.3
11	Burundi	79.2	11	Senegal	45.7
12	Laos	78.5	12	Mauritania	45.9
13	Eritrea	78.4	13	Tunisia	46.1
14	Mozambique	78.1	14	Bosnia & Herz.	46.4
15	Peru	77.6		Egypt	46.4
	Togo	77.6	16	Lebanon	47.0
17	Angola	77.5	17	Papua New Guinea	47.2
18	Vietnam	77.4	18	Somalia	47.4
19	Malawi	76.7	19	Sudan	48.4
20	Liberia	76.3	20	Afghanistan	48.9
21	Cameroon	76.1	21	India	49.3
20	Iceland	75.0	22	Italy	49.6
23	Kenya	74.7	23	Libya	49.7
24	Bahamas	74.6	24	Suriname	51.1
	Zambia	74.6	25	Croatia	51.2
26	Kuwait	73.5	26	Greece	51.8
27	Bahrain	73.4	27	Eswatini	52.5
			28	Pakistan	52.6

Women on boards[b]
OECD countries, %, 2018

1	Iceland	45.9	18	South Africa	27.7
2	France	45.2	19	Spain	26.4
3	Norway	40.2	20	United States	26.1
4	New Zealand	38.2	21	Ireland	26.0
5	Sweden	37.5	22	Switzerland	24.9
6	Italy	36.1	23	Portugal	24.6
7	Belgium	35.9		Slovenia	24.6
8	Germany	35.6	25	Poland	23.5
9	Finland	34.2	26	Israel	21.6
	Netherlands	34.2	27	Czech Republic	18.2
11	United Kingdom	32.6	28	Turkey	18.1
12	Latvia	31.7	29	India	15.9
13	Austria	31.3	30	Colombia	13.5
14	Australia	31.2	31	Luxembourg	13.1
15	Denmark	30.0	32	Hungary	12.9
16	Canada	29.1	33	Lithuania	12.0
	Slovakia	29.1	34	Brazil	11.9

a Aged 15 and over.   b Largest publically listed companies.

Highest rate of unemployment

% of labour force[a], 2019

1	South Africa	28.2	**18**	Spain	14.0
2	West Bank & Gaza	26.2	**19**	Haiti	13.8
3	Lesotho	23.4	**20**	Turkey	13.5
4	Eswatini	22.1	**21**	Yemen	12.9
5	Namibia	20.3	**22**	Iraq	12.8
6	Gabon	20.0		New Caledonia	12.8
7	Libya	18.6	**24**	Serbia	12.7
8	Bosnia & Herz.	18.4	**25**	Albania	12.3
9	Botswana	18.2	**26**	South Sudan	12.2
10	Macedonia	17.8	**27**	Brazil	12.1
11	Greece	17.2		French Polynesia	12.1
12	Armenia	17.0	**29**	Costa Rica	11.9
13	Sudan	16.5		Guyana	11.9
14	Tunisia	16.0	**31**	Algeria	11.7
15	Montenegro	14.9	**32**	Iran	11.4
16	Jordan	14.7		Somalia	11.4
17	Georgia	14.4		Zambia	11.4

Highest rate of youth unemployment

% of labour force[a] aged 15–24, 2019

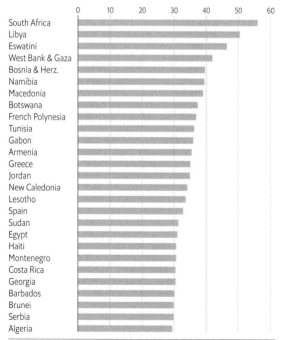

a ILO definition.

Unpaid work[a]
Average minutes per day, 2018 or latest

1	Mexico	270	16	Finland	197
2	Australia	243		United States	197
3	Slovenia	231	18	Germany	196
4	Poland	226		Latvia	196
5	Portugal	223		Norway	196
6	Italy	219		Sweden	196
7	Denmark	217	22	United Kingdom	195
	Spain	217	23	Greece	193
9	Ireland	213		Netherlands	193
10	Belgium	211	25	India	191
11	Estonia	208	26	Canada	186
12	New Zealand	204	27	South Africa	182
13	Austria	202	28	France	181
14	Hungary	200	29	Luxembourg	179
15	Turkey	199	30	China	164

Average hours worked
Per employed person per week, 2018 or latest[b]

1	Mauritania	54.1	12	Turkey	46.2
2	Egypt	53.0	13	Macau	46.0
3	Belize	50.0	14	Guyana	45.8
	Mongolia	50.0	15	Malaysia	45.4
	Qatar	50.0	16	Namibia	44.9
6	Bangladesh	48.6	17	Peru	44.6
7	Myanmar	47.4	18	Nepal	44.5
8	Brunei	47.3	19	Colombia	44.4
9	Sri Lanka	46.8		Uganda	44.4
10	Mexico	46.3	21	Singapore	44.2
	Nigeria	46.3	22	Saudi Arabia	44.1

Gender pay gap[c]
2018 or latest

1	South Korea	34.1	16	Netherlands	14.1
2	Estonia	28.3	17	Mexico	14.0
3	Japan	24.5	18	France	13.0
4	Israel	21.8	19	Chile	12.5
5	Latvia	21.1		Lithuania	12.5
6	United States	18.9	21	Australia	11.7
7	Canada	18.5	22	Iceland	11.5
8	Finland	17.7		Spain	11.5
9	United Kingdom	16.4	24	Hungary	9.4
10	Germany	16.2		Poland	9.4
11	Slovakia	15.7	26	New Zealand	7.9
12	Austria	15.4	27	Sweden	7.3
13	Czech Republic	15.1	28	Norway	7.1
14	Portugal	14.8	29	Turkey	6.9
	Switzerland	14.8	30	Ireland	5.9

a OECD countries.   b 2015–18
c Difference between male and female median wages divided by the male median wages.

Business costs and foreign direct investment

Office rents

Rent, taxes and operating expenses, Q1 2019, $ per sq ft

Hong Kong (Central)	
London (West End), UK	
Hong Kong (Kowloon)	
New York (Midtown Manhattan), US	
Beijing (Finance Street), China	
Beijing (CBD), China	
New York (Midtown-South Manhattan), US	
Tokyo (Marunouchi Otemachi), Japan	
New Delhi (Connaught Place, CBD), India	
London (City), UK	
San Francisco (Downtown), US	
Shanghai (Pudong), China	
San Francisco (Peninsula), US	
Singapore	
Paris, France	
Shanghai (Puxi), China	
Boston (Downtown), US	
New York (Downtown Manhattan), US	
Seoul (CBD), South Korea	
Los Angeles (Suburban), US	

Foreign direct investment[a]

Inflows, $m, 2018

1	United States	251,814
2	China	139,043
3	Hong Kong	115,662
4	Singapore	77,646
5	Netherlands	69,659
6	United Kingdom	64,487
7	Brazil	61,223
8	Australia	60,438
9	Cayman Islands	57,384
10	Spain	43,591
11	India	42,286
12	Canada	39,625
13	France	37,294
14	Mexico	31,604
15	Germany	25,706
16	Italy	24,276
17	Indonesia	21,980
18	Israel	21,803
19	Vietnam	15,500
20	South Korea	14,479
21	Russia	13,332
22	Turkey	12,944

Outflows, $m, 2018

1	Japan	143,161
2	China	129,830
3	France	102,421
4	Hong Kong	85,162
5	Germany	77,076
6	Netherlands	58,983
7	Canada	50,455
8	United Kingdom	49,880
9	Cayman Islands	40,378
10	South Korea	38,917
11	Singapore	37,143
12	Russia	36,445
13	Spain	31,620
14	Switzerland	26,928
15	Saudi Arabia	21,219
16	Italy	20,576
17	Sweden	20,028
18	Taiwan	18,024
19	Thailand	17,714
20	United Arab Emirates	15,079
21	Ireland	13,272
22	India	11,037

Note: CBD is Central Business District.
a Investment in companies in a foreign country.

Business creativity and research

Entrepreneurial activity

Percentage of population aged 18–64 who are either a nascent entrepreneur[a] or owner-manager of a new business, average 2014–19

Highest		Lowest	
1 Senegal	38.6	1 Suriname	2.1
2 Uganda	35.5	2 Pakistan	3.7
3 Botswana	33.0	3 Kosovo	4.0
4 Ecuador	32.8	4 Italy	4.2
5 Angola	31.2	5 Bulgaria	4.5
6 Cameroon	30.1	6 Japan	4.8
7 Burkina Faso	28.3	7 France	5.2
8 Bolivia	27.4	8 Germany	5.4
9 Chile	27.1	9 Denmark	5.5
10 Lebanon	24.9	10 Bosnia & Herz.	5.7
11 Peru	24.6	11 Belarus	5.8
12 Guatemala	22.6	Belgium	5.8
13 Sudan	22.2	13 Spain	5.9
14 Colombia	21.8	14 Finland	6.3
15 Armenia	21.0	Macedonia	6.3
16 Madagascar	20.7	16 Russia	6.5
17 Brazil	19.9	17 Greece	6.6
18 Thailand	19.1	Norway	6.6
19 Philippines	17.8	19 Oman	6.9
20 Vietnam	17.4	Slovenia	6.9
21 Barbados	16.9	21 Sweden	7.3
El Salvador	16.9	22 Morocco	7.4
23 Mexico	16.8	23 Georgia	7.9
24 Canada	16.7	24 Switzerland	8.0
25 Panama	16.0	25 Poland	8.1
26 Turkey	15.2		

Brain gains[b]

Highest, 2019		Lowest, 2019	
1 Singapore	6.1	1 Bosnia & Herz.	1.5
Switzerland	6.1	Venezuela	1.5
3 Luxembourg	5.9	3 Croatia	2.0
4 China	5.6	Macedonia	2.0
United Arab Emirates	5.6	5 Iran	2.2
6 Hong Kong	5.5	Slovakia	2.2
Netherlands	5.5	7 Mauritania	2.3
8 Qatar	5.4	Serbia	2.3
Saudi Arabia	5.4	Zimbabwe	2.3
United States	5.4	10 Bolivia	2.4
11 Bahrain	5.3	Bulgaria	2.4
Canada	5.3	Greece	2.4
13 Guinea	5.2	Haiti	2.4
14 Azerbaijan	5.1	Romania	2.4
New Zealand	5.1	Tunisia	2.4
United Kingdom	5.1	16 Nicaragua	2.5
		Turkey	2.5

a An individual who has started a new firm which has not paid wages for over three months.
b Scores: 1=attracts no highly skilled individuals from abroad; 7=attracts many highly skilled individuals from abroad.

Total spending on R&D

	$bn, 2018				*% of GDP, 2018*	
1	United States	581.6		1	Israel	4.94
2	China	297.4		2	South Korea	4.53
3	Japan	162.3		3	Switzerland[a]	3.37
4	Germany	123.7		4	Taiwan	3.36
5	South Korea	77.9		5	Sweden	3.34
6	France	61.1		6	Japan	3.28
7	United Kingdom	48.7		7	Austria	3.17
8	Italy	29.0		8	Germany	3.09
9	Canada	26.8		9	Denmark	3.03
10	Australia[a]	25.3		10	United States	2.84
11	Switzerland[a]	22.9		11	Belgium	2.76
12	Brazil[b]	22.7			Finland	2.76
13	Taiwan	20.4		13	France	2.20
14	Netherlands	19.8		14	Netherlands	2.17
15	Sweden	18.4		15	China	2.14
16	Israel	18.3		16	Norway	2.07

Innovation index[c]

100=maximum score, 2019

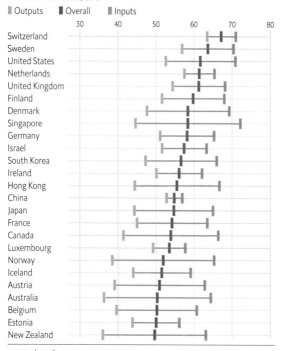

a 2017   b 2016
c The innovation index averages countries' capacity for innovation (inputs) and success
   in innovation (outputs), based on 79 indicators.

Businesses and banks

Largest non-financial companies
By market capitalisation, $trn, end-December 2019

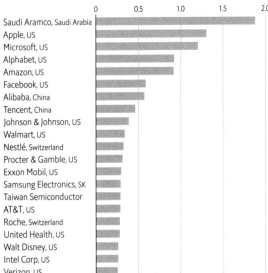

Saudi Aramco, Saudi Arabia	
Apple, US	
Microsoft, US	
Alphabet, US	
Amazon, US	
Facebook, US	
Alibaba, China	
Tencent, China	
Johnson & Johnson, US	
Walmart, US	
Nestlé, Switzerland	
Procter & Gamble, US	
Exxon Mobil, US	
Samsung Electronics, SK	
Taiwan Semiconductor	
AT&T, US	
Roche, Switzerland	
United Health, US	
Walt Disney, US	
Intel Corp, US	
Verizon, US	

By net profit, $bn, 2019

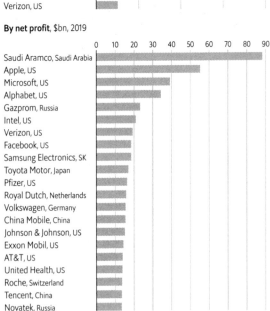

Saudi Aramco, Saudi Arabia	
Apple, US	
Microsoft, US	
Alphabet, US	
Gazprom, Russia	
Intel, US	
Verizon, US	
Facebook, US	
Samsung Electronics, SK	
Toyota Motor, Japan	
Pfizer, US	
Royal Dutch, Netherlands	
Volkswagen, Germany	
China Mobile, China	
Johnson & Johnson, US	
Exxon Mobil, US	
AT&T, US	
United Health, US	
Roche, Switzerland	
Tencent, China	
Novatek, Russia	

Largest banks

By market capitalisation, $bn, end December 2019

1	JPMorgan Chase	United States	437.2
2	Bank of America	United States	316.8
3	Industrial & Commercial Bank of China	China	294.5
4	Wells Fargo	United States	227.5
5	China Construction Bank	China	217.7
6	Agricultural Bank of China	China	182.7
7	Citigroup	United States	174.4
8	HSBC	United Kingdom	159.5
9	Bank of China	China	147.4
10	China Merchants Bank	China	134.9
11	Royal Bank of Canada	Canada	113.3
12	TD Bank Group	Canada	102.1
13	Commonwealth Bank of Australia	Australia	99.3
14	HDFC	India	97.8
15	U.S. Bancorp	United States	92.6

By assets, $bn, end December 2019

1	Industrial & Commercial Bank of China	China	4,308
2	China Construction Bank	China	3,639
3	Agricultural Bank of China	China	3,559
4	Bank of China	China	3,257
5	Mitsubishi UFJ	Japan	2,881
6	HSBC	United Kingdom	2,715
7	JPMorgan Chase	United States	2,687
8	Bank of America	United States	2,434
9	BNP Paribas	France	2,432
10	Crédit Agricole	France	2,260
11	Citigroup	United States	1,951
12	Sumitomo Mitsui Financial	Japan	1,947
13	Wells Fargo	United States	1,928
14	Mizuho	Japan	1,867
15	Banco Santander	Spain	1,711

Largest sovereign-wealth funds

By assets, $bn, April 2020

1	Government Pension Fund, Norway	1,187
2	China Investment Corporation	941
3	Abu Dhabi Investment Authority, UAE	580
4	Hong Kong Monetary Authority Investment Portfolio	540
5	Kuwait Investment Authority	534
6	Government of Singapore Investment Corporation	440
7	SAFE Investment Company, China	418
8	Temasek Holdings, Singapore	375
9	National Council for Social Security Fund, China	325
10	Public Investment Fund of Saudi Arabia	320
11	Qatar Investment Authority	295
12	Investment Corporation of Dubai	239
13	Mubadala Investment Company, UAE	232
14	Turkey Wealth Fund	222

Note: Countries listed refer to the company's domicile.

Stockmarkets

Largest market capitalisation
$bn, end 2019

1	NYSE	23,328
2	Nasdaq – US	13,002
3	Japan Exchange Group	6,191
4	Shanghai SE	5,106
5	Hong Kong Exchanges	4,899
6	Euronext	4,702
7	London SE Group[a]	4,183
8	Shenzhen SE	3,410
9	TMX Group	2,409
10	Saudi SE (Tadawul)	2,407
11	BSE India	2,180
12	National Stock Exchange of India	2,163
13	Deutsche Börse	2,098
14	SIX Swiss Exchange	1,834
15	Nasdaq OMX Nordic Exchanges[b]	1,613
16	Australian Securities Exchange	1,488
17	Korea Exchange[c]	1,485
18	Taiwan SE	1,217
19	Brasil Bolsa Balcão	1,187
20	Johannesburg SE	1,056
21	BME Spanish Exchanges	797
22	Moscow Exchange	792
23	Singapore Exchange	697
24	The Stock Exchange of Thailand	569
25	Indonesia SE	523
26	Bolsa Mexicana de Valores	414
27	Bursa Malaysia	404
28	Tehran SE	321
29	Oslo Bors	296
30	Philippine SE	275
31	Tel-Aviv SE	237
32	Bolsa de Comercio de Santiago	204
33	Borsa Istanbul	185
34	Qatar SE	160
35	Warsaw SE	152
36	Abu Dhabi Securities Exchange	145
37	Ho Chi Minh SE	142
38	Wiener Börse	133
39	Bolsa de Valores de Colombia	132

Stockmarket gains and losses
$ terms, % change December 31st 2018 to December 31st 2019

Best performance

1	Greece (ATHEX comp)	46.8
2	Russia (RTS)	45.3
3	United States (NAS comp)	35.2
4	China (Shenzhen comp)	33.9
5	Israel (TA 125)	31.2
6	United States (S&P 500)	28.9
7	Switzerland (SMI)	28.2
8	Brazil (BVSP)	26.8
9	Taiwan (TWI)	26.5
10	Italy (FTSE MIB)	26.0
11	Canada (S&P TSX)	25.5
12	France (CAC 40)	24.1
13	Denmark (OMXCB)	23.9
14	Germany (DAX)[d]	23.2
15	Euro area (EURO STOXX 50)	22.5
16	United States (DJIA)	22.3
17	Netherlands (AEX)	21.7
18	China (Shanghai comp)	20.5
19	Euro area (FTSE EURO 100)	19.9
20	Belgium (BEL 20)	19.8

Worst performance

1	Chile (IGPA)	-16.8
2	Argentina (MERV)	-13.5
3	Malaysia (KLCI)	-5.1
4	Pakistan (KSE)	-1.5
5	Poland (WIG)	-0.6
6	South Korea (KOSPI)	3.9
7	Indonesia (IDX)	5.3
8	Singapore (STI)	6.5
9	Saudi Arabia (TASI)	7.2
10	Peru (IGBVL)	8.2
11	Mexico (IPC)	9.0
12	Hong Kong (Hang Seng)	9.6
13	Spain (IBEX 35)	9.8
	Thailand (SET)	9.8
15	South Afria (FTSE JSE)	11.4
16	India BSE (SENSEX 30)	11.9
17	Turkey (BIST)	12.1
18	Hungary (BUX)	12.2
19	Czech Republic (SE PX)	12.5
20	Norway (OSEAX)	12.7

a Includes Borsa Italiana.   b Armenia, Copenhagen, Helsinki, Iceland, Riga, Stockholm, Tallinn and Vilnius stock exchanges.   c Includes Kosdaq.   d Total return index.

Value traded[a]

$trn, 2019

Nasdaq – US	
NYSE	
Shenzhen SE	
Shanghai SE	SIX Swiss Exchange
Cboe Global Markets	Brasil Bolsa Balcão
Cboe Europe	Australian Securities Exchange
Japan Exchange Group	
BATS Global Markets – US	Taiwan SE
London SE Group[b]	Nasdaq OMX Nordic Exchanges[d]
Hong Kong Exchanges	
Euronext	BME Spanish Exchanges
Korea Exchange[c]	The Stock Exchange of Thailand
Deutsche Börse	
TMX Group	Johannesburg SE
National Stock Exchange of India	Borsa Istanbul
	Taipei Exchange

Number of listed companies[e]

End 2019

1	BSE India	5,519	21	The Stock Exchange of Thailand	725
2	Japan Exchange Group	3,708	22	Singapore Exchange	723
3	TMX Group	3,413	23	Indonesia SE	668
4	Nasdaq – US	3,140	24	Deutsche Börse	522
5	BME Spanish Exchanges	2,896	25	Tel-Aviv SE	442
6	Hong Kong Exchanges	2,449	26	Borsa Istanbul	379
7	London SE Group[b]	2,410	27	Ho Chi Minh SE	378
8	Korea Exchange[c]	2,283	28	Hanoi SE	367
9	Shenzhen SE	2,205	29	Johannesburg SE	343
10	NYSE	2,143	30	Tehran SE	331
11	Australian Securities Exchange	2,092	31	Brasil Bolsa Balcão	328
12	National Stock Exchange of India	1,955	32	Dhaka SE	320
13	Shanghai SE	1,572	33	Chittagong SE	291
14	Euronext	1,220	34	Colombo SE	289
15	Nasdaq OMX Nordic Exchanges[d]	1,082	35	Bolsa de Comercio de Santiago	283
16	Taiwan SE	956	36	SIX Swiss Exchange	272
17	Bursa Malaysia	927	37	Philippine SE	268
18	Warsaw SE	824	38	Bulgarian SE	262
19	Wiener Börse	778	39	The Egyptian Exchange	247
20	Taipei Exchange	775	40	Moscow Exchange	217

Note: Figures are not entirely comparable due to different reporting rules and calculations.
a Includes electronic and negotiated deals.   b Includes Borsa Italiana.
c Includes Kosdaq.   d Armenia, Copenhagen, Helsinki, Iceland, Riga, Stockholm, Tallinn and Vilnius stock exchanges.   e Domestic and foreign.

Public finance

Government debt
As % of GDP, 2019

1	Japan	224.7	16	Germany	68.5
2	Greece	190.8	17	Poland	65.2
3	Italy	149.5	18	Netherlands	62.8
4	Portugal	136.3	19	Iceland	62.6
5	France	123.0	20	Slovakia	62.5
6	Belgium	117.9	21	Israel	62.1
7	Spain	114.3	22	Sweden	46.9
8	United Kingdom	111.8	23	Norway[a]	45.7
9	United States	108.4	24	Denmark	45.5
10	Canada	95.5	25	Latvia	43.2
11	Austria	94.7	26	Australia	41.5
12	Hungary	84.2	27	Switzerland	40.7
13	Slovenia	82.1	28	Lithuania	40.3
14	Finland	72.0	29	South Korea	38.7
15	Ireland	71.4	30	Czech Republic	38.5

Government spending
As % of GDP, 2019

1	France	54.2	16	Slovakia	40.8
2	Finland	51.6	17	Czech Republic	40.2
3	Belgium	50.9	18	Poland	40.0
4	Denmark	49.1	19	Canada	39.9
5	Norway	48.5	20	United Kingdom	39.1
6	Italy	48.3	21	Iceland	38.8
7	Sweden	47.7	22	Luxembourg	38.6
8	Greece	47.2	23	Israel	38.5
9	Austria	47.0	24	United States	37.6
10	Germany	43.8	25	Japan	36.7
11	Portugal	43.0	26	Estonia	36.1
12	Slovenia	41.6	27	Latvia	35.8
13	Netherlands	41.1	28	Lithuania	33.8
14	Hungary	40.9	29	Australia	33.1
	Spain	40.9	30	New Zealand	32.4

Tax revenue
As % of GDP, 2018

1	France	46.1	14	Hungary	36.6
2	Denmark[b]	44.9	15	Slovenia	36.4
3	Belgium	44.8	16	Portugal	35.4
4	Sweden	43.9	17	Czech Republic	35.3
5	Finland	42.7	18	Poland	35.0
6	Austria	42.2	19	Spain[b]	34.4
7	Italy	42.1	20	United Kingdom	33.5
8	Luxembourg	40.1	21	Estonia	33.2
9	Norway	39.0	22	Slovakia	33.1
10	Netherlands	38.8	23	Canada	33.0
11	Greece	38.7	24	New Zealand	32.7
12	Germany	38.2	25	Japan	31.4
13	Iceland	36.7	26	Israel	31.1

Note: Includes only OECD countries.   a 2018   b 2017

Democracy

Democracy index

Most democratic = 10, 2019

Most			Least		
1	Norway	9.87	**1**	North Korea	1.08
2	Iceland	9.58	**2**	Congo-Kinshasa	1.13
3	Sweden	9.39	**3**	Central African Rep.	1.32
4	New Zealand	9.26	**4**	Syria	1.43
5	Finland	9.25	**5**	Chad	1.61
6	Ireland	9.24	**6**	Turkmenistan	1.72
7	Canada	9.22	**7**	Equatorial Guinea	1.92
	Denmark	9.22	**8**	Saudi Arabia	1.93
9	Australia	9.09		Tajikistan	1.93
10	Switzerland	9.03	**10**	Yemen	1.95
11	Netherlands	9.01	**11**	Uzbekistan	2.01
12	Luxembourg	8.81	**12**	Libya	2.02
13	Germany	8.68	**13**	Laos	2.14
14	United Kingdom	8.52	**14**	Burundi	2.15
15	Uruguay	8.38	**15**	China	2.26
16	Austria	8.29	**16**	Eritrea	2.37
	Spain	8.29	**17**	Iran	2.38
18	Mauritius	8.22	**18**	Belarus	2.48
19	Costa Rica	8.13	**19**	Bahrain	2.55
20	France	8.12	**20**	Guinea-Bissau	2.63

Parliamentary seats

Lower or single house, seats per 100,000 population, March 2020

Most			Fewest		
1	Liechtenstein	65.8	**1**	India	0.04
2	Monaco	61.5	**2**	United States	0.13
3	Andorra	36.4	**3**	Pakistan	0.16
4	Iceland	17.8	**4**	Nigeria	0.18
5	Maldives	16.9	**5**	China	0.21
6	Malta	13.8		Indonesia	0.21
7	Montenegro	13.0	**7**	Bangladesh	0.22
8	Barbados	10.5	**8**	Brazil	0.24
9	Bahamas	10.1	**9**	Iran	0.25
10	Luxembourg	9.9	**10**	Philippines	0.29

Women in parliament

Lower or single house, women as % of total seats, March 2020

1	Rwanda	61.3	**12**	Spain	44.0
2	Cuba	53.2	**13**	Senegal	43.0
3	Bolivia	53.1	**14**	Namibia	42.7
4	United Arab Emirates	50.0	**15**	Switzerland	41.5
5	Mexico	48.2	**16**	Norway	41.4
6	Nicaragua	47.3	**17**	Mozambique	41.2
7	Sweden	47.0	**18**	Argentina	40.9
8	South Africa	46.6	**19**	New Zealand	40.8
9	Andorra	46.4	**20**	Belgium	40.7
10	Finland	46.0	**21**	Belarus	40.0
11	Costa Rica	45.6		Portugal	40.0

Education

Primary enrolment
Number enrolled as % of relevant age group

Highest			Lowest		
1	Madagascar	143	1	Equatorial Guinea	62
2	Malawi	142	2	Eritrea	68
	Nepal	142	3	South Sudan	73
4	Rwanda	133	4	Niger	75
5	Sweden	127	5	Mali	76
6	Namibia	124	6	Sudan	77
	Togo	124	7	Bahamas	81
8	Benin	122		Jordan	81
9	Burundi	121		Senegal	81
	Lesotho	121	10	Puerto Rico	83
11	Bangladesh	116	11	Liberia	85
12	Brazil	115		Nigeria	85
	Colombia	115		Romania	85
	Eswatini	115	14	Chad	87
	Timor-Leste	115	15	Bulgaria	89
	Tunisia	115			

Highest secondary enrolment
Number enrolled as % of relevant age group

1	Belgium	159		Uruguay	120
2	Finland	154	13	Estonia	118
3	Sweden	153		Iceland	118
4	Australia	150		Thailand	118
5	Netherlands	136	16	Liechtenstein	117
6	Costa Rica	133		Norway	117
7	Denmark	129	18	Slovenia	116
8	Spain	126	19	New Zealand	115
	United Kingdom	126	20	Canada	114
10	Ireland	125	21	Kazakhstan	113
11	Portugal	120	22	Latvia	111

Highest tertiary enrolment[a]
Number enrolled as % of relevant age group

1	Greece	137		United States	88
2	Australia	113	12	Belarus	87
3	Puerto Rico	97	13	Austria	85
4	South Korea	94		Netherlands	85
5	Macau	91		Singapore	85
6	Argentina	90	16	New Zealand	82
7	Spain	89		Norway	82
8	Chile	88		Russia	82
	Finland	88	19	Denmark	81
	Latvia	88	20	Belgium	80

Notes: Latest available year 2015–19. The gross enrolment ratios shown are the actual number enrolled as a percentage of the number of children in the official primary age group. They may exceed 100 when children outside the primary age group are receiving primary education.

a Tertiary education includes all levels of post-secondary education including courses leading to awards not equivalent to a university degree, courses leading to a first university degree and postgraduate courses.

Literacy rate
% of adult population[a]

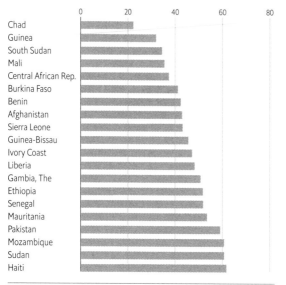

	0	20	40	60	80
Chad					
Guinea					
South Sudan					
Mali					
Central African Rep.					
Burkina Faso					
Benin					
Afghanistan					
Sierra Leone					
Guinea-Bissau					
Ivory Coast					
Liberia					
Gambia, The					
Ethiopia					
Senegal					
Mauritania					
Pakistan					
Mozambique					
Sudan					
Haiti					

Education spending
% of GDP[a]

Highest			Lowest		
1	Norway	8.0	1	South Sudan	1.0
2	Sweden	7.7	2	Bermuda	1.5
3	Iceland	7.5		Congo-Kinshasa	1.5
4	Sierra Leone	7.1		Monaco	1.5
5	Costa Rica	7.0	5	Papua New Guinea	1.9
6	Finland	6.9	6	Bangladesh	2.0
7	Tunisia	6.6		Myanmar	2.0
8	Belgium	6.5	8	Sri Lanka	2.1
	Lesotho	6.5	9	Cambodia	2.2
10	New Zealand	6.4		Chad	2.2
11	Cyprus	6.3	11	Bahrain	2.3
12	Brazil	6.2	12	Gambia, The	2.4
	South Africa	6.2	13	Albania	2.5
14	Honduras	6.1		Azerbaijan	2.5
15	Burkina Faso	6.0		Uganda	2.5
	Kyrgyzstan	6.0	16	Guinea	2.6
17	Guyana	5.9		Liberia	2.6
18	Israel	5.8		Mauritania	2.6
19	Czech Republic	5.6	19	Armenia	2.7
	Mozambique	5.6		Macau	2.7

a Latest year 2014–18.

Marriage and divorce

Highest marriage rates
Number of marriages per 1,000 population, 2018 or latest

1	West Bank & Gaza	10.0		Latvia	6.8
2	Fiji	9.8		Singapore	6.8
3	Egypt	9.6	23	Barbados	6.7
4	Bahamas	9.5		Bermuda	6.7
	Uzbekistan	9.5	25	Hong Kong	6.6
6	Tajikistan	8.9		Mongolia	6.6
7	Albania	8.0	27	Macedonia	6.5
8	Mauritius	7.9	28	Azerbaijan	6.4
9	Kyrgyzstan	7.8		Belarus	6.4
	Sri Lanka	7.8		Jamaica	6.4
11	Guam	7.5	31	Malta	6.3
	Iran	7.5	32	Georgia	6.2
	Kazakhstan	7.5		Israel	6.2
	Moldova	7.5	34	Liechtenstein	6.0
15	Romania	7.3	35	Brunei	5.9
16	Lithuania	7.1		Malaysia	5.9
	Turkey	7.1		Ukraine	5.9
18	Jordan	6.9	38	Macau	5.8
	United States	6.9	39	Slovakia	5.7
20	Cyprus	6.8			

Lowest marriage rates
Number of marriages per 1,000 population, 2018 or latest

1	Qatar	1.4	25	Mexico	4.0
2	French Guiana	2.4	26	Bulgaria	4.1
3	Peru	2.5		Norway	4.1
4	Venezuela	2.6	28	Guatemala	4.2
5	Guadeloupe	2.8	29	Finland	4.3
	Martinique	2.8		New Zealand	4.3
	Uruguay	2.8		Puerto Rico	4.3
8	Argentina	2.9	32	Greece	4.4
	Isle of Man	2.9		Ireland	4.4
10	Italy	3.2		United Kingdom	4.4
	Luxembourg	3.2	35	French Polynesia	4.5
12	Chile	3.3	36	Australia	4.6
	New Caledonia	3.3		Curaçao	4.6
	Panama	3.3	38	Costa Rica	4.7
	Réunion	3.3	39	Japan	4.8
16	Portugal	3.4		Monaco	4.8
	Suriname	3.4		Switzerland	4.8
18	France	3.5	42	Croatia	4.9
	Kuwait	3.5		Dominican Rep.	4.9
	Slovenia	3.5	44	Estonia	5.0
	Spain	3.5		Germany	5.0
22	Netherlands	3.7		South Korea	5.0
23	Andorra	3.9		Sweden	5.0
	Belgium	3.9			

Note: The data are based on latest available figures (no earlier than 2014) and hence will be affected by the population age structure at the time. Marriage rates refer to registered marriages only and, therefore, reflect the customs surrounding registry and efficiency of administration.

Highest divorce rates
Number of divorces per 1,000 population, 2018 or latest[a]

1	Moldova	4.0	15	Sweden	2.5
2	Belarus	3.5		United States	2.5
3	Guam	3.4	17	Ecuador	2.4
4	Lithuania	3.1		Estonia	2.4
5	Kazakhstan	3.0		Finland	2.4
	Puerto Rico	3.0		Liechtenstein	2.4
	Ukraine	3.0	21	Czech Republic	2.3
8	Latvia	2.9		Macau	2.3
9	Georgia	2.8	23	Cyprus	2.2
10	Bermuda	2.7		Iran	2.2
	Cuba	2.7	25	Egypt	2.1
12	Costa Rica	2.6		Jordan	2.1
	Curaçao	2.6		Portugal	2.1
	Denmark	2.6		South Korea	2.1

Lowest divorce rates
Number of divorces per 1,000 population, 2018 or latest[a]

1	Guatemala	0.3	15	Jamaica	1.2
2	Qatar	0.4		Mexico	1.2
3	Peru	0.5	17	Armenia	1.3
4	Bosnia & Herz.	0.6		Mongolia	1.3
5	Ireland	0.7	19	Montenegro	1.4
	Malta	0.7		Serbia	1.4
	Venezuela	0.7	21	Azerbaijan	1.5
8	French Guiana	0.8		Brunei	1.5
	Macedonia	0.8		Bulgaria	1.5
10	Martinique	1.0		Croatia	1.5
	Uzbekistan	1.0		Italy	1.5
12	Panama	1.1		New Zealand	1.5
	Slovenia	1.1		Suriname	1.5
	Tajikistan	1.1			

Mean age of women at first marriage
Years, 2016 or latest[b]

Youngest			Oldest		
1	Barbados	17.0	1	Slovenia	34.1
2	Niger	17.2	2	French Polynesia	33.8
3	Central African Rep.	17.3	3	Estonia	33.6
4	South Sudan	18.4	4	Hungary	32.9
5	Mozambique	18.7	5	New Caledonia	32.8
6	Bangladesh	18.8	6	Lithuania	32.7
	Chad	18.8	7	Bulgaria	32.6
8	Mali	19.1	8	Czech Republic	32.4
9	Guyana	19.2		Iceland	32.4
10	Guinea	19.8		Ireland	32.4
11	Burkina Faso	20.0		Netherlands	32.4
12	Nepal	20.2	12	Italy	32.2
13	Malawi	20.4	13	Norway	32.1
14	Equatorial Guinea	20.5			

a No earlier than 2014.   b No earlier than 2000.

Households, living costs and giving

Number of households
Biggest, m, 2018 or latest

1	China	473.8	21	Turkey	23.6
2	India	282.2	22	Egypt	23.5
3	United States	138.5	23	South Korea	20.4
4	Indonesia	66.9	24	Spain	18.6
5	Brazil	64.8	25	South Africa	18.0
6	Russia	57.2	26	Ukraine	16.7
7	Japan	54.2	27	Poland	14.6
8	Germany	40.8	28	Colombia	14.4
9	Nigeria	40.6	29	Argentina	14.3
10	Mexico	33.8	30	Canada	14.1
11	Bangladesh	33.3	31	Kenya	11.2
12	France	29.8	32	Australia	9.2
13	Pakistan	29.4	33	Taiwan	8.6
14	United Kingdom	29.0		Venezuela	8.6
15	Vietnam	28.9	35	Peru	8.4
16	Italy	25.9	36	Algeria	8.1
17	Iran	24.7	37	Morocco	7.9
18	Philippines	23.9	38	Netherlands	7.8
19	Thailand	23.8	39	Malaysia	7.7
20	Ethiopia	23.7	40	Romania	7.5

Average household size, people

Biggest, 2018 or latest			*Smallest, 2018 or latest*		
1	Senegal	8.7	1	Germany	2.0
2	Gambia, The	8.2	2	Finland	2.1
3	Afghanistan	8.0	3	Denmark	2.2
4	Congo-Kinshasa	7.3		Estonia	2.2
5	Pakistan	7.2		France	2.2
6	Guinea	6.7		Latvia	2.2
	Yemen	6.7		Netherlands	2.2
8	Bahrain	6.1		Norway	2.2
9	Tajikistan	6.0		Sweden	2.2
10	Burkina Faso	5.9	10	Austria	2.3
	Niger	5.9		Italy	2.3
	Sierra Leone	5.9		Japan	2.3
13	Chad	5.8		Lithuania	2.3
	Mali	5.8		Slovenia	2.3
	United Arab Emirates	5.8		Switzerland	2.3
16	Saudi Arabia	5.7		United Kingdom	2.3
17	Maldives	5.4	17	Belarus	2.4
	Uzbekistan	5.4		Belgium	2.4
19	Timor-Leste	5.3		Czech Republic	2.4
20	Algeria	5.2		Hungary	2.4
	Benin	5.2		United States	2.4
	Uganda	5.2			

a The cost of living index shown is compiled by the Economist Intelligence Unit for use by companies in determining expatriate compensation: it is a comparison of the cost of maintaining a typical international lifestyle in the country rather than a comparison of the purchasing power of a citizen of the country. The index is based on typical urban prices an international executive and family will face abroad. The prices

Cost of living[a]

December 2019, US=100

Highest

Hong Kong	
Singapore	
France	
Israel	
Japan	
Switzerland	
Denmark	
Norway	
South Korea	
Austria	
Australia	
Finland	
Jordan	
Ireland	
United Kingdom	
Thailand	
Iceland	
New Caledonia	
Canada	
New Zealand	
Spain	
Belgium	
Italy	
Netherlands	
UAE	
China	

Lowest

Syria	
Uzbekistan	
Kazakhstan	
Argentina	
Pakistan	
Venezuela	
Zambia	
Algeria	
India	
Nigeria	
Romania	
Paraguay	
Sri Lanka	
Nepal	
Panama	
Hungary	
Senegal	
South Africa	
Bulgaria	
Egypt	
Iran	
Morocco	
Saudi Arabia	
Ivory Coast	
Oman	
Serbia	

World Giving Index[b]

Top givers, % of population, 2019[c]

1	Myanmar	58	16	Denmark	44
	United States	58		Liberia	44
3	New Zealand	57	18	Germany	43
4	Australia	56		Turkmenistan	43
	Ireland	56	20	Nigeria	42
6	Canada	55		Sierra Leone	42
7	United Kingdom	54		Thailand	42
8	Netherlands	53	23	Cyprus	41
9	Sri Lanka	51		Finland	41
10	Indonesia	50		Iran	41
11	Kenya	47		Uzbekistan	41
	Malta	47	27	Guatemala	40
13	Austria	45		Haiti	40
	Switzerland	45		Luxembourg	40
	United Arab Emirates	45		Sweden	40

are for products of international comparable quality found in a supermarket or department store. Prices found in local markets and bazaars are not used unless the available merchandise is of the specified quality and the shopping area itself is safe for executive and family members. New York City prices are used as the base, so United States = 100.
b Three criteria are used to assess giving: in the previous month those surveyed either gave money to charity, gave time to those in need or helped a stranger.
c Ten year average.

Transport: roads and cars

Longest road networks
Km, 2018 or latest

1	United States	6,853,024	26	Argentina	230,923
2	India	5,903,293	27	Vietnam	228,645
3	China	4,846,500	28	Colombia	219,159
4	Brazil	1,751,868	29	Philippines	217,456
5	Russia	1,452,200	30	Malaysia	216,604
6	Canada	1,409,008	31	Hungary	206,632
7	Japan	1,226,559	32	Nigeria	198,233
8	France	1,079,398	33	Peru	173,209
9	South Africa	891,932	34	Egypt	167,774
10	Australia	872,848	35	Ukraine	163,028
11	Spain	683,175	36	Kenya	161,415
12	Germany	644,480	37	Belgium	155,357
13	Indonesia	537,838	38	Congo-Kinshasa	154,633
14	Italy	494,844	39	Netherlands	139,448
15	Finland	454,000	40	Austria	133,597
16	Sweden	432,009	41	Czech Republic	130,671
17	Poland	423,997	42	Algeria	119,386
18	United Kingdom	422,310	43	Greece	116,986
19	Thailand	396,721	44	Sri Lanka	114,093
20	Turkey	394,390	45	Ethiopia	110,414
21	Mexico	393,473	46	Ghana	109,515
22	Iran	312,761	47	South Korea	105,758
23	Saudi Arabia	274,861	48	Belarus	99,606
24	Bangladesh	266,216	49	Zimbabwe	97,724
25	Pakistan	263,775	50	Kazakhstan	96,643

Densest road networks
Km of road per sq km land area, 2018 or latest

1	Monaco	38.5		Sri Lanka	1.8
2	Macau	13.8		Switzerland	1.8
3	Malta	9.7	26	Czech Republic	1.7
4	Bermuda	8.3		Italy	1.7
5	Bahrain	5.3		United Kingdom	1.7
6	Belgium	5.1	29	Austria	1.6
7	Singapore	4.9	30	Finland	1.5
8	Netherlands	4.1	31	Cyprus	1.4
9	Barbados	3.7		Estonia	1.4
10	Japan	3.4		Ireland	1.4
11	Puerto Rico	3.0		Poland	1.4
12	Liechtenstein	2.5		Spain	1.4
13	Hungary	2.3	36	Lithuania	1.3
14	Luxembourg	2.2	37	Latvia	1.1
	Slovenia	2.2		Mauritius	1.1
16	Bangladesh	2.0		South Korea	1.1
	France	2.0		Sweden	1.1
	Hong Kong	2.0		Taiwan	1.1
	India	2.0	42	Greece	0.9
	Jamaica	2.0		Israel	0.9
21	Guam	1.9		Portugal	0.9
22	Denmark	1.8		Qatar	0.9
	Germany	1.8			

Most crowded road networks

Number of vehicles per km of road network, 2018 or latest

1	United Arab Emirates	587.4	26	Tunisia	77.6
2	Monaco	427.3	27	Dominican Rep.	77.2
3	Hong Kong	321.1	28	Germany	75.3
4	Kuwait	302.0	29	Romania	74.5
5	Macau	283.1	30	Iraq	72.2
6	Singapore	248.3	31	Barbados	71.7
7	South Korea	209.4	32	Switzerland	71.0
8	Taiwan	190.8	33	Portugal	69.5
9	Jordan	175.2	34	Netherlands	69.2
10	Israel	167.9	35	Malaysia	67.3
11	Bahrain	154.8	36	Argentina	66.7
12	Puerto Rico	133.4	37	Morocco	66.5
13	Guatemala	125.6	38	Slovakia	65.7
14	Mauritius	121.6	39	Chile	63.6
15	Brunei	118.3	40	Japan	63.5
16	Guam	117.5	41	Armenia	62.8
17	Malta	109.9	42	Poland	61.9
18	Qatar	109.8	43	Moldova	61.3
19	Syria	109.1	44	Ukraine	59.6
20	Mexico	103.2	45	Georgia	57.8
21	Lebanon	100.0	46	Ecuador	56.8
22	Bulgaria	94.5	47	Croatia	56.6
23	United Kingdom	92.1	48	Greece	54.4
24	Italy	86.5	49	Bermuda	53.2
25	Luxembourg	84.3	50	Kazakhstan	52.1

Most road deaths

Fatalities per 100,000 population, 2016

1	Liberia	35.9	24	Chad	27.6
2	Burundi	34.7	25	Benin	27.5
	Zimbabwe	34.7	26	Congo-Brazzaville	27.4
4	Congo-Kinshasa	33.7	27	Somalia	27.1
	Venezuela	33.7	28	Ethiopia	26.7
6	Central African Rep.	33.6	29	Syria	26.5
7	Thailand	32.7	30	Vietnam	26.4
8	Guinea-Bissau	31.1	31	Niger	26.2
9	Malawi	31.0	32	Libya	26.1
10	Burkina Faso	30.5	33	South Africa	25.9
11	Namibia	30.4	34	Sudan	25.7
12	Cameroon	30.1	35	Eritrea	25.3
	Mozambique	30.1	36	Ghana	24.9
14	South Sudan	29.9	37	Mauritania	24.7
15	Rwanda	29.7	38	Equatorial Guinea	24.6
16	Tanzania	29.2		Guyana	24.6
	Togo	29.2	40	Jordan	24.4
18	Uganda	29.0	41	Botswana	23.8
19	Lesotho	28.9	42	Angola	23.6
20	Saudi Arabia	28.8		Ivory Coast	23.6
21	Madagascar	28.6		Malaysia	23.6
22	Guinea	28.2	45	Senegal	23.4
23	Kenya	27.8	46	Gabon	23.2

Highest car ownership
Number of cars per 1,000 population, 2018

1	Puerto Rico	862	26	Sweden	476
2	Brunei	768	27	Guam	470
3	New Zealand	685	28	Bulgaria	466
4	Iceland	651	29	Portugal	451
5	Luxembourg	646	30	Denmark	426
6	Italy	622	31	Cyprus	425
7	Canada	612	32	Lithuania	424
8	Malta	591	33	Ireland	420
9	Poland	580	34	Malaysia	409
10	Australia	560	35	Kuwait	405
11	Austria	545	36	Slovakia	395
12	Germany	542	37	United States	365
13	Switzerland	540	38	Croatia	361
14	Estonia	533	39	Libya	350
15	Slovenia	531	40	Bahrain	342
16	United Kingdom	504	41	Belarus	341
17	Norway	503	42	South Korea	339
18	Netherlands	500	43	Latvia	336
19	Czech Republic	499	44	Hungary	332
20	Belgium	497	45	Israel	328
	France	497	46	Russia	327
22	Greece	495	47	Barbados	321
23	Japan	485	48	Syria	294
24	Spain	483	49	Bermuda	289
25	Finland	479	50	Taiwan	284

Lowest car ownership
Number of cars per 1,000 population, 2018

1	Ethiopia	0.9		Kenya	18.1
2	Sudan	1.3	24	India	18.4
3	Burundi	1.9		Togo	18.4
4	Bangladesh	2.5	26	Nicaragua	18.9
5	Malawi	3.5	27	Benin	19.2
6	Haiti	3.7	28	Cuba	20.8
7	Uganda	3.8		Ghana	20.8
8	Honduras	4.5	30	Yemen	22.4
9	Liberia	4.6	31	Senegal	23.4
10	Tanzania	4.8		Vietnam	23.4
11	Mauritania	5.0	33	Angola	26.2
12	Madagascar	8.3	34	Bolivia	26.6
13	Mali	9.7	35	Philippines	31.5
14	Cameroon	10.9	36	Sri Lanka	33.2
15	Burkina Faso	11.0	37	Afghanistan	35.1
16	Mozambique	11.1	38	Guatemala	43.2
17	Congo-Kinshasa	13.5	39	Egypt	47.7
18	Pakistan	13.8	40	Iraq	48.1
19	Zambia	14.2	41	Peru	49.4
20	Nigeria	16.4	42	Jamaica	52.5
21	El Salvador	18.1	43	Indonesia	55.4
	Ivory Coast	18.1	44	Paraguay	60.0

Car production

Number of cars produced, m, 2018

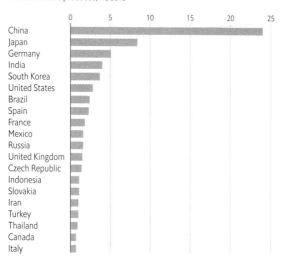

Cars sold

New car registrations, '000, 2018

1	China	23,710		26	South Africa	365
2	United States	5,304		27	Sweden	354
3	Japan	4,391		28	Austria	341
4	Germany	3,436		29	Saudi Arabia	340
5	India	3,395		30	Chile	315
6	United Kingdom	2,367		31	Switzerland	299
7	France	2,173		32	Czech Republic	261
8	Brazil	2,102		33	Israel	257
9	Italy	1,910		34	Philippines	253
10	Russia	1,607		35	Portugal	228
11	South Korea	1,525		36	Colombia	220
12	Spain	1,321		37	Denmark	219
13	Iran	913		38	Pakistan	217
14	Mexico	883		39	Vietnam	198
15	Indonesia	879		40	Morocco	163
16	Australia	874		41	Norway	148
17	Thailand	730		42	Egypt	146
18	Argentina	611		43	Hungary	137
19	Canada	578		44	Romania	129
20	Belgium	550		45	Ireland	126
21	Malaysia	533		46	Algeria	124
22	Poland	532		47	Finland	120
23	Turkey	486		48	Peru	115
24	Netherlands	444		49	New Zealand	108
25	Taiwan	389		50	Ecuador	105

Transport: planes and trains

Most air travel
Passengers carried, m, 2018

1	United States	889.0		16	Thailand	76.1
2	China	611.4		17	Australia	75.7
3	Ireland	167.6		18	France	70.2
4	United Kingdom	165.4		19	Mexico	64.6
5	India	164.0		20	Malaysia	60.5
6	Japan	126.4		21	Hong Kong	47.1
7	Turkey	115.6		22	Vietnam	47.0
8	Indonesia	115.2		23	Netherlands	44.0
9	Germany	109.8		24	Philippines	43.1
10	Brazil	102.1		25	Singapore	40.4
11	Russia	99.3		26	Saudi Arabia	39.1
12	United Arab Emirates	95.5		27	Colombia	33.7
13	Canada	89.4		28	Hungary	31.2
14	South Korea	88.2		29	Qatar	29.2
15	Spain	80.7		30	Switzerland	28.9

Busiest airports
Total passengers, m, 2019

1	Atlanta, Hartsfield	110.5
2	Beijing, Capital	100.0
3	Los Angeles, Intl.	88.1
4	Tokyo, Haneda	87.1
5	Dubai Intl.	86.4
6	Chicago, O'Hare	84.6
7	London, Heathrow	80.8
8	Paris, Charles de Gaulle	76.2
	Shanghai, Pudong Intl.	76.2
10	Dallas, Ft Worth	75.1
11	Guangzhou Baiyun, Intl	73.4
12	Amsterdam, Schiphol	71.7
13	Hong Kong, Intl.	71.5
14	Seoul, Incheon	71.2
15	Frankfurt, Main	70.6

Total cargo, m tonnes, 2019

1	Hong Kong, Intl.	4.8
	Memphis, Intl.	4.8
3	Shanghai, Pudong Intl.	3.6
4	Louisville, Muhammad Ali Intl.	2.9
5	Seoul, Incheon	2.8
6	Dubai, Intl.	2.5
7	Miami, Intl.	2.3
8	Paris, Charles de Gaulle	2.2
	Taiwan, Taoyuan Intl.	2.2
10	Frankfurt, Main	2.1
11	Beijing, Capital	2.0
	Singapore, Changi	2.0
	Tokyo, Narita	2.0
14	Guangzhou Baiyun, Intl	1.9

Average daily aircraft movements, take-offs and landings, 2019

1	Chicago, O'Hare	2,520		16	London, Heathrow	1,304
2	Atlanta, Hartsfield	2,478		17	San Francisco, Intl.	1,256
3	Dallas, Ft Worth	1,973		18	New York, JFK	1,249
4	Los Angeles, Intl.	1,894		19	Indonesia, Soekarno-Hatta Intl.	1,226
5	Denver, Intl.	1,754		20	Phoenix, Skyharbor Intl.	1,202
6	Beijing, Capital	1,628		21	Mexico City, Intl.	1,185
7	Charlotte/Douglas, Intl.	1,584		22	Madrid Barajas	1,168
8	Las Vegas, McCarran Intl.	1,515		23	Hong Kong, Intl.	1,151
9	Frankfurt, Main	1,408		24	Munich	1,143
10	Shanghai, Pudong Intl.	1,402		25	Miami, Intl.	1,142
11	Paris, Charles de Gaulle	1,365		26	Minneapolis–Saint Paul	1,113
12	Amsterdam, Schiphol	1,361		27	Seoul, Incheon	1,107
13	Indira Gandhi, Intl.	1,353		28	Kuala Lumpur Intl.	1,095
14	Guangzhou Baiyun, Intl	1,346		29	Detroit Metro	1,087
15	Houston, George Bush Intl.	1,310				

Longest railway networks

'000 km, 2018 or latest

#	Country		#	Country	
1	United States	150.5	21	Sweden	9.7
2	Russia	85.6	22	Czech Republic	9.4
3	India	68.4	23	Iran	9.3
4	China	67.5	24	Indonesia	8.4
5	Canada	47.7	25	Pakistan	7.8
6	Germany	33.4	26	Turkmenistan	7.7
7	Australia	32.8	27	Hungary	7.1
8	France	28.2	28	Finland	5.9
9	Ukraine	21.6	29	Belarus	5.5
10	South Africa	21.0	30	Egypt	5.2
11	Poland	18.5	31	Austria	4.9
12	Argentina	17.6	32	Uzbekistan	4.6
13	Japan	16.9	33	Thailand	4.5
14	Italy	16.8	34	Sudan	4.3
15	Kazakhstan	16.1	35	South Korea	4.2
16	United Kingdom	16.0	36	Norway	4.1
17	Spain	15.6	37	Algeria	4.0
18	Mexico	14.4		Bulgaria	4.0
19	Romania	10.8	39	Serbia	3.7
20	Turkey	10.3	40	Congo-Kinshasa	3.6

Most rail passengers

Km per person per year, 2018 or latest

#	Country		#	Country	
1	Switzerland	2,106	13	Finland	773
2	Japan	1,551	14	Czech Republic	729
3	France	1,435	15	Luxembourg	729
4	Austria	1,352	16	Slovakia	700
5	Netherlands	1,086	17	Belarus	657
6	Kazakhstan	1,050	18	Italy	651
7	United Kingdom	1,048	19	Sweden	642
8	Denmark	1,013	20	Ukraine	633
9	Germany	956	21	Spain	598
10	Belgium	933	22	Norway	580
11	Russia	888	23	Hungary	568
12	India	850	24	China	477

Most rail freight

Million tonne-km per year, 2018 or latest

#	Country		#	Country	
1	United States	2,525,217	13	Japan	21,265
2	Russia	2,491,876	14	Lithuania	16,885
3	China	2,238,435	15	France	14,842
4	India	620,175	16	Mongolia	13,493
5	Kazakhstan	206,258	17	Turkmenistan	13,327
6	Ukraine	191,914	18	Latvia	12,186
7	Germany	70,614	19	Turkey	12,058
8	Australia	59,649	20	Czech Republic	11,819
9	Belarus	52,574	21	Finland	11,030
10	Austria	32,406	22	Italy	9,478
11	Iran	30,299	23	Switzerland	8,256
12	Uzbekistan	22,940	24	Pakistan	8,080

Transport: shipping

Merchant fleets

Number of vessels, by country of domicile, January 2020

1	China	6,125	16	India	1,019
2	Greece	4,536	17	Taiwan	1,005
3	Japan	3,822	18	Denmark	980
4	Singapore	2,727	19	United Arab Emirates	913
5	Germany	2,672	20	France	883
6	Indonesia	2,145	21	Italy	692
7	Norway	2,038	22	Malaysia	599
8	United States	1,978	23	Bermuda	532
9	Russia	1,707	24	Philippines	488
10	South Korea	1,647	25	Switzerland	437
11	Hong Kong	1,628	26	Thailand	406
12	Turkey	1,522	27	Brazil	401
13	United Kingdom	1,336	28	Canada	373
14	Netherlands	1,195	29	Ukraine	350
15	Vietnam	1,020	30	Cyprus	300

Ships' flags

Registered fleet, 2019

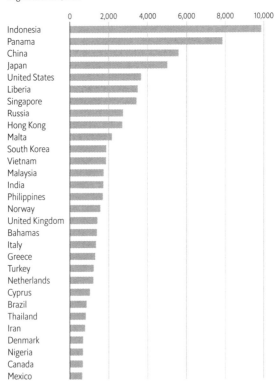

Crime and punishment

Murders

Homicides per 100,000 pop., 2017 or latest

1	El Salvador	61.8
2	Jamaica	57.0
3	Venezuela	56.3
4	Virgin Islands (US)	49.3
5	Honduras	41.7
6	Lesotho	41.3
7	South Africa	35.9
8	Bahamas	30.9
	Trinidad & Tobago	30.9
10	Brazil	30.5
11	Guatemala	26.1
12	Colombia	24.9
13	Mexico	24.8
14	Central African Rep.	19.8
15	Curaçao	19.2
16	Puerto Rico	18.5
17	Namibia	17.1
18	Botswana	15.0
19	Guyana	14.8
20	South Sudan	13.9

Robberies

Per 100,000 pop., 2017 or latest

1	Costa Rica	1,587
2	Argentina	920
3	Brazil	797
4	Chile	633
5	Uruguay	522
6	Ecuador	457
7	South Africa	332
8	Guatemala	263
9	Colombia	244
10	Panama	243
11	Peru	223
12	Trinidad & Tobago	182
13	Mexico	179
14	Maldives	175
15	Belgium	167
16	Bahamas	166
17	Guyana	160
18	France	150
19	Dominican Rep.	144
20	Spain	142

Prisoners

Total prison pop., 2020 or latest

1	United States	2,121,600
2	China	1,710,000
3	Brazil	773,151
4	Russia	517,028
5	India	466,084
6	Thailand	373,169
7	Turkey	286,000
8	Iran	240,000
9	Indonesia	224,522
10	Philippines	215,000
11	Mexico	198,384
12	South Africa	154,437
13	Vietnam	123,697
14	Colombia	118,079
15	Ethiopia	113,727
16	Egypt	106,000
17	Argentina	103,209
18	Peru	94,235
19	Myanmar	92,000
20	Bangladesh	88,084
21	Morocco	85,767
22	United Kingdom	80,827
23	Pakistan	77,275
24	Poland	74,154
25	Malaysia	74,000
26	Nigeria	71,938
27	France	70,651

Per 100,000 pop., 2020 or latest

1	United States	655
2	El Salvador	590
3	Turkmenistan	552
4	Thailand	538
5	Rwanda	511
6	Cuba	510
7	Maldives	499
8	Bahamas	442
9	Panama	419
10	Guam	411
11	Virgin Islands (US)	394
12	Costa Rica	374
13	Brazil	366
14	Cayman Islands	365
15	Russia	358
16	Turkey	344
17	Belarus	343
18	Bermuda	342
19	Uruguay	337
20	Nicaragua	332
21	Puerto Rico	313
22	Brunei	307
23	Barbados	300
24	Namibia	295
25	Iran	294
26	Trinidad & Tobago	292
27	Peru	286

War and terrorism

Defence spending
As % of GDP, 2019

1	Oman	11.7		Cambodia	3.9
2	Afghanistan	10.2	14	Azerbaijan	3.8
3	Saudi Arabia	10.1		Iran	3.8
4	Iraq	9.1	16	Pakistan	3.7
5	Algeria	6.0	17	Brunei	3.5
6	Israel	5.8	18	Bulgaria	3.3
7	Armenia	4.8	19	Colombia	3.2
8	Kuwait	4.7		United States	3.2
9	Jordan	4.6	21	Morocco	3.1
10	Mali	4.1		Myanmar	3.1
11	Trinidad & Tobago	4.0		Singapore	3.1
12	Bahrain	3.9			

Defence spending

$bn, 2019			Per person, $, 2019		
1	United States	684.6	1	Oman	2,517
2	China	181.1	2	Saudi Arabia	2,331
3	Saudi Arabia	78.4	3	Israel	2,254
4	Russia[a]	61.6	4	Kuwait	2,165
5	India	60.5	5	United States	2,063
6	United Kingdom	54.8	6	Singapore	1,846
7	France	52.3	7	Norway	1,240
8	Japan	48.6	8	Australia	1,074
9	Germany	48.5	9	Bahrain	1,018
10	South Korea	39.8	10	Brunei	950
11	Brazil	27.5	11	United Kingdom	837
12	Italy	27.1	12	Denmark	786
13	Australia	25.5	13	France	773
14	Israel[b]	22.6	14	South Korea	770
15	Iraq	20.5	15	Trinidad & Tobago	746

Armed forces
'000, 2019

		Regulars	Reserves			Regulars	Reserves
1	China	2,035	510	16	Colombia	293	35
2	India	1,456	1,155	17	Sri Lanka	255	6
3	United States	1,380	849	18	Japan	247	56
4	North Korea	1,280	600	19	Mexico	236	82
5	Russia	900	2,000	20	Saudi Arabia	227	0
6	Pakistan	654	0	21	Ukraine	209	900
7	Iran	610	350	22	France	204	39
8	South Korea	599	3,100	23	Eritrea	202	120
9	Vietnam	482	5,000	24	Morocco	196	150
10	Egypt	439	479	25	Iraq	193	0
11	Myanmar	406	0	26	South Sudan	185	0
12	Indonesia	396	400	27	Germany	181	29
13	Brazil	367	1,340	28	Afghanistan	181	0
14	Thailand	361	200	29	Israel	170	465
15	Turkey	355	379	30	Syria	169	0

a National defence budget only.   b Includes US Foreign Military Assistance.

Arms exporters

$m, 2019

1	United States	10,752
2	Russia	4,718
3	France	3,368
4	China	1,423
5	Germany	1,185
6	Spain	1,061
7	United Kingdom	972
8	South Korea	688
9	Italy	491
10	Israel	369
11	Netherlands	285
12	Switzerland	254
13	Turkey	245
14	Sweden	206
15	Canada	188

Arms importers

$m, 2019

1	Saudi Arabia	3,673
2	India	2,964
3	Qatar	2,258
4	South Korea	1,510
5	Australia	1,399
6	Egypt	1,193
7	United States	1,048
8	Japan	891
9	China	887
10	Turkey	833
11	Bangladesh	743
12	UAE	644
13	Singapore	614
14	Pakistan	561
15	Israel	507

Terrorist attacks

Number of incidents, 2018

Country	Per m pop.
Afghanistan	47.7
Iraq	35.5
India	0.7
Nigeria	3.3
Philippines	5.6
Somalia	35.1
Pakistan	2.3
Yemen	11.4
Cameroon	9.3
Syria	13.7
Colombia	4.1
Thailand	2.6
Libya	24.8
Mali	8.6
Congo-Kinshasa	1.9
West Bank & Gaza	23.9
United Kingdom	1.5
Nepal	3.5
Turkey	1.1
Saudi Arabia	2.7
Burkina Faso	3.5
United States	0.2
Israel	7.5
Mozambique	2.1
Egypt	0.5
Kenya	1.0
Central African Rep.	9.8
Chile	2.4
Indonesia	0.2
Sudan	1.0

Space and peace

Orbital launches

2019	Low-earth orbit[a]	Greater than low-earth orbit	Total
1 China	21	13	34
2 Russia	15	10	25
3 United States	11	8	19
4 Europe	2	4	6
5 India	5	1	6
6 New Zealand	6	0	6
7 Iran	2	0	2
Japan	2	0	2
Ukraine	2	0	2

Satellites in space

By country of ownership[b], 2019

1 United States	1,308	Italy	12
2 China	356	Netherlands	12
3 Russia	167	19 Brazil	11
4 United Kingdom	130	Taiwan	11
5 Japan	78	21 Indonesia	9
6 India	58	22 Singapore	8
7 Canada	39	United Arab Emirates	8
8 Germany	37	24 Finland	7
9 Luxembourg	32	Kazakhstan	7
10 Spain	21	Norway	7
11 Argentina	16	Turkey	7
South Korea	16	28 Algeria	5
13 Israel	15	Denmark	5
14 Saudi Arabia	13	Malaysia	5
15 Australia	12	South Africa	5
France	12	Thailand	5

Global Peace Index[c]

Most peaceful, 2019		Least peaceful, 2019	
1 Iceland	1.078	1 Afghanistan	3.644
2 New Zealand	1.198	2 Syria	3.539
3 Portugal	1.247	3 Iraq	3.487
4 Austria	1.275	4 South Sudan	3.447
5 Denmark	1.283	5 Yemen	3.411
6 Canada	1.298	6 Somalia	3.302
7 Singapore	1.321	7 Libya	3.258
8 Czech Republic	1.337	8 Congo Kinshasa	3.243
9 Japan	1.360	9 Central African Rep.	3.237
10 Switzerland	1.366	10 Russia	3.049
11 Slovenia	1.369	11 Sudan	3.043
12 Ireland	1.375	12 Pakistan	2.973
13 Australia	1.386	13 North Korea	2.962
14 Finland	1.404	14 Turkey	2.959
15 Sweden	1.479	15 Venezuela	2.936
16 Germany	1.494	16 Ukraine	2.927
		17 Nigeria	2.865

a Up to 2,000km.   b Excludes multi-country/agency.
c Ranks 163 countries using 23 indicators which gauge the level of safety and security in society, the extent of domestic or international conflict and the degree of militarisation.

Environment

Biggest emitters of carbon dioxide
Million tonnes, 2016

1	China	10,501.8	21	Taiwan	329.7
2	United States	5,174.8	22	Poland	327.3
3	India	2,237.1	23	Spain	310.3
4	Russia	1,689.8	24	United Arab Emirates	289.2
5	Japan	1,138.8	25	Kazakhstan	279.6
6	Germany	834.7	26	Singapore	247.3
7	Iran	630.7	27	Ukraine	233.1
8	Saudi Arabia	611.5	28	Egypt	226.3
9	South Korea	610.7	29	Malaysia	217.2
10	Canada	541.4	30	Argentina	213.8
11	Brazil	521.6	31	Vietnam	211.8
12	Indonesia	500.5	32	Pakistan	176.0
13	United Kingdom	474.9	33	Netherlands	168.1
14	South Africa	474.5	34	Algeria	136.9
15	France	411.4	35	Venezuela	132.2
16	Mexico	410.6	36	Iraq	122.2
17	Australia	408.5	37	Belgium	116.6
18	Turkey	395.6	38	Qatar	116.5
19	Italy	382.6	39	Philippines	116.4
20	Thailand	343.9	40	Czech Republic	110.6

Largest amount of carbon dioxide emitted per person
Tonnes, 2016

1	Singapore	44.1	13	Australia	16.8
2	Qatar	43.9	14	Malta	16.3
3	Trinidad & Tobago	35.5	15	United States	16.0
4	United Arab Emirates	30.8	16	Kazakhstan	15.7
5	Bahrain	29.0	17	Oman	15.2
6	Kuwait	26.0	18	Canada	15.0
7	Virgin Islands (US)	25.3	19	Taiwan	14.0
8	Brunei	23.1	20	Hong Kong	13.2
9	Luxembourg	22.3	21	Bermuda	12.8
10	Saudi Arabia	18.8	22	South Korea	12.0
11	New Caledonia	18.5	23	Russia	11.6
12	Turkmenistan	17.8	24	Czech Republic	10.5

Change in carbon emissions
Million tonnes, 2016 compared with 2006

Biggest increases			Biggest decreases		
1	China	3,761.8	1	United States	-736.0
2	India	976.2	2	Ukraine	-128.8
3	Saudi Arabia	205.2	3	United Kingdom	-119.1
4	Indonesia	175.3	4	Japan	-97.8
5	Iran	153.2	5	Italy	-87.4
6	Turkey	142.9	6	Netherlands	-84.0
7	United Arab Emirates	132.9	7	Spain	-66.6
8	Brazil	127.5	8	North Korea	-49.6
9	South Korea	122.4	9	Mexico	-39.7
10	Vietnam	119.4	10	Greece	-36.7
11	Thailand	112.1	11	Canada	-36.3

Worst air pollution

Particulate matter concentration[a], micrograms per cubic metre
Average for urban areas, 2016

1	Nepal	99.5	26	Turkey	41.2
2	Qatar	91.7	27	Eritrea	41.1
3	Saudi Arabia	86.7	28	South Sudan	40.9
4	Egypt	79.6	29	Rwanda	40.7
5	Niger	73.0	30	Senegal	39.7
6	Bahrain	69.0	31	Gabon	37.8
7	India	68.0	32	Congo-Kinshasa	37.4
8	Cameroon	65.4		Syria	37.4
9	Iraq	60.1	34	United Arab Emirates	37.2
10	Afghanistan	59.9	35	Congo-Brazzaville	36.4
11	Kuwait	58.9	36	Burkina Faso	36.3
12	Bangladesh	58.6	37	Oman	36.2
13	Pakistan	56.2	38	Tunisia	35.7
14	Central African Rep.	51.2	39	Burundi	35.6
15	China	51.0	40	Myanmar	34.6
16	Chad	50.8	41	Algeria	34.5
17	Mongolia	49.5	42	Iran	34.4
18	Equatorial Guinea	49.1	43	Ethiopia	34.0
19	Uganda	48.7	44	Macedonia	33.0
20	Sudan	46.8	45	Armenia	32.9
21	Nigeria	46.3	46	Gambia, The	32.3
22	Yemen	44.3	47	Jordan	31.7
23	Tajikistan	42.8	48	Togo	31.2
24	Libya	41.7	49	Ghana	31.1
	Mauritania	41.7		Morocco	31.1

Least access to drinking water[b]

% of population, 2017

1	Chad	38.7	21	Burundi	60.8
2	South Sudan	40.7		Sierra Leone	60.8
3	Ethiopia	41.1	23	Guinea	61.9
4	Papua New Guinea	41.3	24	Yemen	63.5
5	Congo-Kinshasa	43.2	25	Zimbabwe	64.1
6	Central African Rep.[c]	46.3	26	Equatorial Guinea	64.7
7	Burkina Faso	47.9	27	Togo	65.1
8	Uganda	49.1	28	Haiti	65.5
9	Niger	50.3	29	Benin	66.4
10	Eritrea[c]	51.8	30	Guinea-Bissau	66.6
11	Somalia	52.4	31	Afghanistan	67.1
12	Madagascar	54.4	32	Lesotho	68.6
13	Mozambique	55.7	33	Malawi	68.8
14	Angola	55.8	34	Eswatini	69.0
15	Tanzania	56.7	35	Mauritania	70.7
16	Rwanda	57.7	36	Nigeria	71.4
17	Kenya	58.9	37	Ivory Coast	72.9
18	Zambia	60.0		Liberia	72.9
19	Sudan	60.3	39	Congo-Brazzaville	73.2
20	Cameroon	60.4	40	Gambia, The	78.0

a Particulates less than 2.5 microns in diameter.   b From an improved source.   c 2016

Open defecation[a]

Prevalence among households, %, 2016

1	Niger	68.8	15	India	28.4
2	Chad	67.3	16	Lesotho	28.3
3	Eritrea	67.0	17	Ivory Coast	26.3
4	South Sudan	64.3	18	Ethiopia	25.6
5	Benin	54.1	19	Zimbabwe	25.3
6	Namibia	49.1	20	Sudan	24.4
7	Togo	48.4	21	Nepal	24.0
8	Burkina Faso	48.0	22	Central African Rep.	23.8
9	Madagascar	44.2	23	Laos	22.9
10	Liberia	40.5	24	Angola	21.2
11	Cambodia	34.5	25	Timor-Leste	20.9
12	Mauritania	33.2	26	Haiti	20.7
13	Mozambique	29.3	27	Nigeria	20.2
14	Somalia	29.2	28	Yemen	19.8

Handwashing

Lowest access to basic facilities, such as soap and water, 2017

	% of urban population			% of rural population	
1	Liberia	1.8	1	Liberia	0.6
2	Lesotho	5.9	2	Lesotho	0.7
3	Congo-Kinshasa	7.4	3	Gambia, The	1.1
4	Guinea-Bissau	8.8	4	Congo-Kinshasa	2.2
5	Somalia	12.0	5	Chad	2.3
6	Gambia, The	12.3	6	Rwanda	2.8
7	Rwanda	13.4	7	Cameroon	2.9
8	Cameroon	14.6	8	Togo	3.7
9	Malawi	15.3	9	Ethiopia	4.0
10	Benin	16.7	10	Burundi	4.1
11	Chad	17.7	11	Guinea-Bissau	4.6
12	Togo	20.1	12	Zambia	5.2
13	Burundi	20.3	13	Benin	6.0
14	Burkina Faso	22.7	14	Malawi	7.4

Deaths from poor sanitation[b]

Per 100,000 pop., 2016

1	Chad	101.0	14	Angola	48.8
2	Somalia	86.6	15	Ivory Coast	47.2
3	Central African Rep.	82.1	16	Eritrea	45.6
4	Sierra Leone	81.3	17	Cameroon	45.2
5	Niger	70.8	18	Guinea	44.6
6	Mali	70.7	19	Lesotho	44.4
7	Nigeria	68.6	20	Ethiopia	43.7
8	Burundi	65.4	21	Togo	41.6
9	South Sudan	63.3	22	Liberia	41.5
10	Congo-Kinshasa	59.8	23	Congo-Brazzaville	38.7
11	Benin	59.7	24	Mauritania	38.6
12	Kenya	51.2	25	Tanzania	38.4
13	Burkina Faso	49.6	26	Guinea-Bissau	35.3

a Includes in fields, bushes, open water, beaches.
b Attributed to unsafe water, unsafe sanitation and lack of hygiene.

EPI ecosystem vitality[a]

100=best, 2018

Top			Bottom		
1	Switzerland	83.3	1	Haiti	26.1
2	France	76.1	2	Bosnia & Herz.	27.2
3	Slovakia	75.1	3	Madagascar	28.0
4	Taiwan	74.8	4	Libya	28.5
5	Austria	74.0	5	Burundi	28.6
6	Malta	72.3	6	Iraq	31.0
7	Azerbaijan	71.5	7	Niger	31.6
	Germany	71.5	8	Mauritania	32.1
9	Sweden	71.2	9	Papua New Guinea	34.1
10	Italy	71.0	10	Eritrea	34.8
11	Denmark	70.5	11	Oman	34.9
12	Hungary	69.9	12	Barbados	36.5
13	Belgium	69.4		Lesotho	36.5
14	United Kingdom	69.1	14	Congo-Kinshasa	37.6
15	Romania	68.9	15	Maldives	38.1
16	Luxembourg	68.5	16	Afghanistan	38.4
17	Spain	67.9	17	Gambia, The	38.7
18	Poland	67.7	18	Guyana	38.9
19	Equatorial Guinea	67.6	19	Benin	39.4
20	Ireland	67.3	20	Angola	39.9

Forests

Biggest % change in forested areas, 2006–16

Deforestation			Reforestation		
1	Togo	-54.1	1	Burundi	43.6
2	Uganda	-41.0	2	Iceland	33.8
3	Nigeria	-38.4	3	Bahrain	32.6
4	Kyrgyzstan	-24.3	4	Azerbaijan	29.0
5	Pakistan	-23.1	5	Montenegro	24.1
6	Honduras	-21.2	6	Rwanda	22.6
7	Chad	-21.0	7	Algeria	21.8
8	North Korea	-20.5	8	Dominican Rep.	19.7
9	Zimbabwe	-18.8	9	Uruguay	19.4
10	Paraguay	-17.5	10	Cuba	18.6
11	Mauritania	-15.6	11	Philippines	17.8
12	Mali	-14.6	12	Lesotho	15.7
13	El Salvador	-14.4	13	French Polynesia	14.8
14	Timor-Leste	-14.2	14	Tunisia	13.0
15	Myanmar	-13.7	15	Moldova	12.5
16	Cambodia	-12.0	16	Chile	12.2
17	Somalia	-10.9		Vietnam	12.2
18	Cameroon	-10.6	18	Laos	11.1
19	Benin	-10.5	19	Costa Rica	10.8
20	Argentina	-10.2		Sierra Leone	10.8
	Burkina Faso	-10.2	21	Kuwait	9.8
22	Guatemala	-10.0	22	Turkey	9.7
23	Niger	-9.9	23	Denmark	9.5

a Assessed on biodiversity, forest loss, fisheries, climate and energy, air pollution, water resources.

Protected areas[a] as % of total land

2017

Highest			*Lowest*		
1	New Caledonia	54.4	**1**	Afghanistan	0.1
2	Venezuela	54.1	**2**	Libya	0.2
3	Slovenia	53.6		Turkey	0.2
4	Brunei	46.9	**4**	Lesotho	0.3
5	Hong Kong	41.9	**5**	Mauritania	0.6
6	Luxembourg	40.9	**6**	Syria	0.7
7	Congo-Brazzaville	40.7	**7**	Yemen	0.8
8	Poland	39.7	**8**	Maldives	1.2
9	Croatia	38.3	**9**	Barbados	1.3
10	Tanzania	38.1	**10**	Bosnia & Herz.	1.4
11	Namibia	37.9	**11**	Iraq	1.5
	Zambia	37.9	**12**	Jordan	1.8
13	Germany	37.8	**13**	Haiti	1.9
14	Slovakia	37.6	**14**	French Polynesia	2.0
15	Nicaragua	37.2	**15**	Bermuda	2.1
16	Bahamas	36.6	**16**	Sudan	2.3
17	Guinea	35.6	**17**	North Korea	2.4
18	Greece	35.2	**18**	Lebanon	2.6
19	Bulgaria	34.7		Oman	2.6
20	Monaco	33.2	**20**	Papua New Guinea	3.1
21	New Zealand	32.5	**21**	Turkmenistan	3.2
22	Bolivia	30.9	**22**	Kazakhstan	3.3
23	Morocco	30.8	**23**	Uruguay	3.4
24	Trinidad & Tobago	30.6		Uzbekistan	3.4
25	Malta	30.3	**25**	Isle of Man	3.7

Protected marine areas[b] as % of territorial waters

2017

1	Slovenia	100.0	**20**	Estonia	18.6
2	Monaco	99.8	**21**	Dominican Rep.	18.0
3	New Caledonia	96.6	**22**	Denmark	17.8
4	Germany	45.4	**23**	Colombia	17.1
5	France	45.0	**24**	Portugal	16.6
6	United States	41.1	**25**	Latvia	16.0
7	Australia	40.6		Sudan	16.0
8	Belgium	36.7	**27**	Sweden	15.2
9	Jordan	35.6	**28**	Ecuador	13.3
10	New Zealand	30.4	**29**	South Africa	12.1
11	United Kingdom	28.9	**30**	United Arab Emirates	11.3
12	Chile	28.8	**31**	Finland	10.5
	Gabon	28.8	**32**	Guinea-Bissau	10.0
14	Netherlands	26.7	**33**	Italy	8.8
15	Brazil	26.6	**34**	Croatia	8.5
16	Lithuania	25.6	**35**	Spain	8.4
17	Romania	23.1	**36**	Japan	8.2
18	Poland	22.6	**37**	Bulgaria	8.1
19	Mexico	21.8	**38**	Bahamas	7.9

a Includes national parks, nature reserves, wildlife sanctuaries.
b For environmental purposes.

Life expectancy

Highest life expectancy

Years, 2020–25

1	Monaco[a]	89.3	25	Guadeloupe	82.7
2	Hong Kong	85.3		Portugal	82.7
3	Japan	85.0	27	Finland	82.5
4	Macau	84.7	28	Belgium	82.2
5	Switzerland	84.3		Liechtenstein[a]	82.2
6	Singapore	84.1	30	Austria	82.1
7	Italy	84.0		Channel Islands	82.1
	Spain	84.0	32	Germany	81.9
9	Australia	83.9		Slovenia	81.9
10	Iceland	83.5	34	United Kingdom	81.8
	Israel	83.5	35	Bermuda[a]	81.7
	South Korea	83.5	36	Cayman Islands[a]	81.6
13	Sweden	83.3		Isle of Man[a]	81.6
14	France	83.1		Réunion	81.6
	Malta	83.1	39	Cyprus	81.5
	Martinique	83.1	40	Denmark	81.4
17	Andorra[a]	83.0	41	Virgin Islands (US)	81.2
	Canada	83.0	42	Taiwan	81.0
19	Norway	82.9	43	Costa Rica	80.9
20	Greece	82.8	44	Chile	80.7
	Ireland	82.8		Guam	80.7
	Luxembourg	82.8		Qatar	80.7
	Netherlands	82.8		Puerto Rico	80.7
	New Zealand	82.8	48	French Guiana	80.5

Highest male life expectancy

Years, 2020–25

1	Monaco[a]	85.4		Sweden	81.7
2	Hong Kong	82.4	12	Malta	81.4
	Switzerland	82.4	13	Ireland	81.3
4	Iceland	82.2		Spain	81.3
5	Australia	82.1	15	Canada	81.2
	Singapore	82.1		Netherlands	81.2
7	Israel	82.0		New Zealand	81.2
8	Italy	81.9	18	Norway	81.1
	Japan	81.9	19	Andorra[a]	80.8
10	Macau	81.7		Luxembourg	80.8

Highest female life expectancy

Years, 2020–25

1	Monaco[a]	93.3	9	Italy	86.0
2	Hong Kong	88.2		Switzerland	86.0
3	Japan	88.1	11	Guadeloupe	85.9
4	Macau	87.6	12	Australia	85.8
5	Spain	86.7		France	85.8
6	South Korea	86.4	14	Andorra[a]	85.4
7	Singapore	86.2	15	Portugal	85.3
8	Martinique	86.1	16	Finland	85.2

a 2019 estimate.

Lowest life expectancy

Years, 2020–25

1	Central African Rep.	54.4	26	Zambia	64.7
2	Chad	55.2	27	Ghana	64.9
3	Lesotho	55.6		Namibia	64.9
4	Nigeria	55.8		South Africa	64.9
5	Sierra Leone	55.9	30	Haiti	65.0
6	Somalia	58.3		Liberia	65.0
7	South Sudan	58.7	32	Congo-Brazzaville	65.2
8	Ivory Coast	58.8		Papua New Guinea	65.2
9	Guinea-Bissau	59.4	34	Malawi	65.6
10	Equatorial Guinea	59.8		Mauritania	65.6
11	Cameroon	60.3	36	Afghanistan	66.0
12	Mali	60.5	37	Sudan	66.1
13	Eswatini	61.0	38	Tanzania	66.4
14	Congo-Kinshasa	61.6		Yemen	66.4
15	Mozambique	62.1	40	Gabon	67.0
	Togo	62.1	41	Eritrea	67.5
17	Angola	62.2		Kenya	67.5
	Zimbabwe	62.2	43	Ethiopia	67.8
19	Guinea	62.6		Myanmar	67.8
20	Burundi	62.7		Pakistan	67.8
21	Benin	62.8	46	Fiji	67.9
22	Burkina Faso	63.0	47	Madagascar	68.2
23	Gambia, The	63.3	48	Turkmenistan	68.6
24	Niger	63.6	49	Laos	68.9
25	Uganda	64.4		Senegal	68.9

Lowest male life expectancy

Years, 2020–25

1	Central African Rep.	52.2	10	Ivory Coast	57.5
2	Lesotho	52.5	11	Equatorial Guinea	58.8
3	Chad	53.7	12	Cameroon	59.0
4	Nigeria	54.8	13	Mozambique	59.1
5	Sierra Leone	55.0	14	Angola	59.5
6	Somalia	56.6	15	Mali	59.7
7	Eswatini	57.0	16	Congo-Kinshasa	60.0
8	South Sudan	57.2	17	Zimbabwe	60.4
9	Guinea-Bissau	57.3	18	Burundi	60.8

Lowest female life expectancy

Years, 2020–25

1	Central African Rep.	56.6	11	Mali	61.4
2	Chad	56.7	12	Cameroon	61.7
3	Nigeria	56.8	13	Togo	63.1
	Sierra Leone	56.8	14	Congo-Kinshasa	63.2
5	Lesotho	58.9	15	Guinea	63.3
6	Ivory Coast	60.1	16	Zimbabwe	63.7
	Somalia	60.1	17	Burkina Faso	63.8
8	South Sudan	60.3	18	Benin	64.5
9	Equatorial Guinea	61.1	19	Burundi	64.6
10	Guinea-Bissau	61.3	20	Gambia, The	64.7

Death rates and infant mortality

Highest death rates
Number of deaths per 1,000 population, 2020–25

#	Country	Rate	#	Country	Rate
1	Bulgaria	15.6	49	Curaçao	9.4
2	Ukraine	15.2		Ivory Coast	9.4
3	Latvia	15.0		South Africa	9.4
4	Lithuania	14.4	52	Barbados	9.3
5	Lesotho	13.4	53	Netherlands	9.2
	Romania	13.4		United States	9.2
7	Croatia	13.3	55	Bermuda[a]	9.1
	Serbia	13.3		Mauritius	9.1
9	Russia	13.1		Sweden	9.1
10	Hungary	13.0	58	Malta	9.0
11	Georgia	12.7		Trinidad & Tobago	9.0
12	Belarus	12.6	60	Eswatini	8.9
13	Estonia	12.2		Guinea-Bissau	8.9
14	Moldova	12.0	62	Congo-Kinshasa	8.8
15	Germany	11.7		Guadeloupe	8.8
16	Bosnia & Herz.	11.5	64	Albania	8.7
	Japan	11.5	65	Fiji	8.6
18	Greece	11.4		Mali	8.6
19	Central African Rep.	11.3	67	Cameroon	8.5
	Chad	11.3	68	Equatorial Guinea	8.4
21	Portugal	11.1		Myanmar	8.4
22	Nigeria	11.0	70	Haiti	8.3
23	Czech Republic	10.9		Thailand	8.3
	Italy	10.9	72	Benin	8.2
25	Monaco[a]	10.8		Switzerland	8.2
	Montenegro	10.8	74	Taiwan	8.1
	Sierra Leone	10.8	75	Canada	7.9
28	Poland	10.7		Guyana	7.9
29	Macedonia	10.5		Norway	7.9
	Slovenia	10.5		Togo	7.9
31	Isle of Man[a]	10.4	79	China	7.8
	Slovakia	10.4		Liechtenstein[a]	7.8
33	Puerto Rico	10.3	81	Andorra[a]	7.7
34	Somalia	10.2		Guinea	7.7
35	Finland	10.1		Jamaica	7.7
36	Austria	10.0	84	Argentina	7.6
	Denmark	10.0		Mozambique	7.6
38	South Sudan	9.9		Suriname	7.6
39	Belgium	9.8		Zimbabwe	7.6
40	Armenia	9.7	88	Angola	7.4
	Cuba	9.7		Cyprus	7.4
	Martinique	9.7		India	7.4
43	North Korea	9.6		Namibia	7.4
	Virgin Islands (US)	9.6		Niger	7.4
45	France	9.5	93	Burundi	7.3
	Spain	9.5		Kazakhstan	7.3
	United Kingdom	9.5		Venezuela	7.3
	Uruguay	9.5			

Note: Both death and, in particular, infant mortality rates can be underestimated in certain countries where not all deaths are officially recorded.   a 2019 estimate.

Highest infant mortality
Number of deaths per 1,000 live births, 2020–25

1	Central African Rep.	71		Mozambique	45
2	Sierra Leone	70	24	Togo	43
3	Chad	67	25	Yemen	42
4	Somalia	63	26	Turkmenistan	40
5	Congo-Kinshasa	60	27	Niger	39
6	South Sudan	59		Uganda	39
7	Equatorial Guinea	58		Zambia	39
8	Mali	57	30	Gambia, The	38
9	Pakistan	56		Sudan	38
10	Nigeria	55	32	Burundi	37
11	Benin	54		Papua New Guinea	37
	Cameroon	54	34	Eswatini	35
	Ivory Coast	54	35	Tanzania	34
14	Angola	53		Zimbabwe	34
15	Guinea-Bissau	50	37	Laos	33
	Lesotho	50		Malawi	33
17	Haiti	48		Myanmar	33
	Mauritania	48	40	Timor-Leste	32
19	Liberia	47	41	Congo-Brazzaville	31
20	Afghanistan	45		Gabon	31
	Burkina Faso	45		Ghana	31
	Guinea	45		Kenya	31

Lowest death rates
No. of deaths per 1,000 pop., 2020–25

1	Qatar	1.5
2	United Arab Emirates	1.8
3	Oman	2.5
4	Bahrain	2.6
5	Maldives	2.7
6	French Guiana	3.1
7	Kuwait	3.3
8	West Bank & Gaza	3.5
9	Saudi Arabia	3.7
10	Jordan	4.0
11	Syria	4.1
12	Macau	4.3
13	Honduras	4.5
14	Tajikistan	4.6
15	Algeria	4.7
	Guatemala	4.7
	Iraq	4.7
18	Iran	4.8
19	Lebanon	4.9
	Rwanda	4.9
21	Brunei	5.0
22	Morocco	5.1
	Nicaragua	5.1

Lowest infant mortality
No. of deaths per 1,000 live births, 2020–25

1	Finland	1
	Hong Kong	1
	Iceland	1
	Singapore	1
5	Austria	2
	Belarus	2
	Belgium	2
	Czech Republic	2
	Estonia	2
	Germany	2
	Greece	2
	Ireland	2
	Israel	2
	Italy	2
	Japan	2
	Luxembourg	2
	Macau	2
	Monaco[a]	2
	Montenegro	2
	Netherlands	2
	Norway	2
	Portugal	2
	Slovenia	2
	South Korea	2
	Spain	2
	Sweden	2

Death and disease

Diabetes

Prevalence in pop. aged 20–79, %,
2019 age-standardised estimate[a]

1	Sudan	22.1
2	Mauritius	22.0
3	New Caledonia	21.8
4	Pakistan	19.9
5	French Polynesia	19.5
6	Guam	18.7
7	Papua New Guinea	17.9
8	Egypt	17.2
9	Malaysia	16.7
10	United Arab Emirates	16.3
11	Saudi Arabia	15.8
12	Bahrain	15.6
	Qatar	15.6
14	Fiji	14.7
15	Puerto Rico	13.7

Ischaemic heart disease

Deaths per 100,000 pop., 2017

1	Ukraine	693
2	Belarus	555
3	Lithuania	540
4	Bulgaria	479
5	Latvia	458
6	Moldova	409
7	Georgia	407
8	Russia	385
9	Serbia	353
10	Hungary	351
11	Romania	348
12	Estonia	331
13	Armenia	312
	Slovakia	312
15	Croatia	304
16	Czech Republic	300
17	Azerbaijan	267

Cancer

Deaths per 100,000 pop., 2016 or latest
available year, age standardised estimate[a]

1	Mongolia	224
2	Hungary	172
3	Croatia	157
4	Slovakia	154
5	Serbia	153
6	Poland	146
7	Latvia	144
8	Romania	143
9	Lithuania	141
	Slovenia	141
	Uruguay	141
12	Estonia	139
13	Armenia	138
14	Denmark	131
15	Cuba	129
	Czech Republic	129
	Netherlands	129
18	Ireland	128
	Moldova	128
	Russia	128
21	Bulgaria	126
22	Macedonia	124
23	United Kingdom	123
24	Barbados	122
25	New Zealand	121

Pollution

Deaths attributable to ambient air
pollution per 100,000 pop., 2016,
age-standardised estimate[a]

1	Sierra Leone	324
2	Nigeria	307
3	Chad	280
4	Ivory Coast	269
5	Niger	252
6	Togo	250
7	Guinea	243
8	Gambia, The	237
9	Guinea-Bissau	215
10	Somalia	213
11	Central African Rep.	212
12	Afghanistan	211
13	Mali	209
14	Cameroon	208
15	North Korea	207
16	Burkina Faso	206
17	Benin	205
18	Ghana	204
19	Nepal	194
	Yemen	194
21	Laos	189
22	Philippines	185
	Sudan	185

a Assumes that every country and region has the same age profile.
Note: Statistics are not available for all countries. The number of cases diagnosed and reported depends on the quality of medical practice and administration and can be under-reported in a number of countries.

Measles immunisation

Lowest % of children aged 12–23 months, 2018

1	Dominican Rep.	31
2	Angola	35
3	Bolivia	38
	Haiti	38
5	Afghanistan	39
	Suriname	39
	Venezuela	39
8	Philippines	40
9	Kenya	45
10	Yemen	46
11	Niger	48
12	Namibia	50
	South Africa	50
14	Syria	54
	Timor-Leste	54
16	Sierra Leone	55
17	Laos	57

DPT[a] immunisation

Lowest % of children aged 12–23 months, 2018

1	Equatorial Guinea	25
2	Chad	41
3	Somalia	42
4	Guinea	45
5	Central African Rep.	47
	Syria	47
7	South Sudan	49
8	Ukraine	50
9	Nigeria	57
10	Angola	59
11	Venezuela	60
12	Papua New Guinea	61
13	Haiti	64
14	Philippines	65
	Yemen	65
16	Afghanistan	66
17	Laos	68

HIV/AIDS prevalence

Prevalence in adults aged 15–49, %, 2018

1	Eswatini	27.3
2	Lesotho	23.6
3	South Africa	20.4
4	Botswana	20.3
5	Zimbabwe	12.7
6	Mozambique	12.6
7	Namibia	11.8
8	Zambia	11.3
9	Malawi	9.2
10	Equatorial Guinea	7.1
11	Uganda	5.7
12	Kenya	4.7
13	Tanzania	4.6
14	Gabon	3.8
15	Cameroon	3.6
	Central African Rep.	3.6
17	Guinea-Bissau	3.5
18	Congo-Brazzaville	2.6
	Ivory Coast	2.6
20	Rwanda	2.5
	South Sudan	2.5
22	Togo	2.3
23	Angola	2.0
	Haiti	2.0
25	Gambia, The	1.9
	Jamaica	1.9
27	Bahamas	1.8

AIDS

Deaths per 100,000 population, 2018

1	Lesotho	289.4
2	Botswana	213.0
3	Eswatini	211.3
4	Mozambique	183.1
5	Zimbabwe	152.4
6	Equatorial Guinea	137.5
7	South Africa	122.9
8	Namibia	110.3
9	Central African Rep.	102.9
10	Zambia	98.0
11	Guinea-Bissau	96.1
12	South Sudan	90.2
13	Congo-Brazzaville	76.3
14	Malawi	71.7
15	Cameroon	71.4
16	Ivory Coast	63.8
17	Gabon	56.6
18	Uganda	53.8
19	Bahamas	51.8
20	Jamaica	51.1
21	Kenya	48.6
22	Togo	48.2
23	Mauritius	48.1
24	Ghana	47.0
25	Angola	45.4
26	Gambia, The	43.0
27	Tanzania	42.6

a Diphtheria, pertussis and tetanus.

Health

Highest health spending
As % of GDP, 2017

1	United States	17.1
2	Sierra Leone	13.4
3	Switzerland	12.3
4	Afghanistan	11.8
5	Cuba	11.7
6	France	11.3
7	Germany	11.2
8	Sweden	11.0
9	Japan	10.9
10	Canada	10.6
11	Armenia	10.4
	Austria	10.4
	Norway	10.4
14	Andorra	10.3
	Belgium	10.3
16	Denmark	10.1
	Netherlands	10.1
18	South Sudan	9.8
19	Malawi	9.6
	United Kingdom	9.6
21	Brazil	9.5
22	Malta	9.3
	Uruguay	9.3

Lowest health spending
As % of GDP, 2017

1	Venezuela	1.2
2	Monaco	1.8
3	Bangladesh	2.3
4	Brunei	2.4
5	Laos	2.5
	Papua New Guinea	2.5
7	Qatar	2.6
8	Angola	2.8
	Gabon	2.8
10	Congo-Brazzaville	2.9
	Eritrea	2.9
	Pakistan	2.9
13	Indonesia	3.0
14	Equatorial Guinea	3.1
	Kazakhstan	3.1
16	Gambia, The	3.3
	Ghana	3.3
	United Arab Emirates	3.3
19	Ethiopia	3.5
	Fiji	3.5
	India	3.5
22	Tanzania	3.6

Out-of-pocket spending
$ per person at PPP, 2016

Highest

1	Switzerland	2,333
2	Andorra	2,086
3	Singapore	1,313
4	Malta	1,244
5	United States	1,105
6	Austria	1,043
7	Cyprus	1,039
8	Azerbaijan	942
9	South Korea	916
10	Trinidad & Tobago	889
11	Norway	877
12	Belgium	871
13	Turkmenistan	852
14	Australia	850
15	Finland	844
16	Sweden	832
17	Italy	794
18	Spain	789
19	Greece	781
20	Portugal	779
21	Bulgaria	767
22	Iceland	727
23	Iran	724
24	Germany	717

Lowest

1	Mozambique	5
2	Rwanda	8
3	Gambia, The	11
	Papua New Guinea	11
5	Central African Rep.	13
	Malawi	13
7	Congo-Kinshasa	14
8	Burundi	15
9	Madagascar	20
10	Zambia	21
11	Ethiopia	23
12	Tanzania	24
13	Timor-Leste	28
14	Mali	29
15	Botswana	31
16	Eritrea	32
17	Benin	36
	Niger	36
19	Burkina Faso	39
20	Kenya	40
21	Lesotho	42
	Zimbabwe	42
23	Uganda	47

Obesity[a]

Children aged 5–9 years as % of population group, 2016

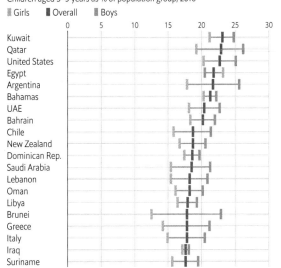

Underweight[b]

Children aged 5–9 years as % of population group, 2016

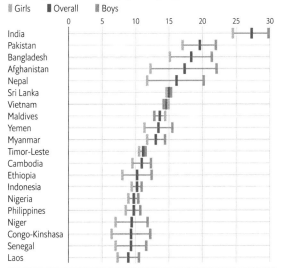

a Body mass index > +2 standard deviations above the median for age, crude estimate.
b Body mass index < -2 standard deviations above the median for age, crude estimate.

Telephones and the internet

Mobile telephones
Subscribers per 100 population, 2018

1	Macau	345.3		26	Malta	140.2
2	Hong Kong	270.0		27	Gambia, The	139.5
3	United Arab Emirates	208.5			Nepal	139.5
4	Montenegro	180.7		29	Cyprus	138.9
5	Thailand	180.2		30	Kyrgyzstan	138.6
6	Kuwait	171.6		31	Gabon	138.3
7	Costa Rica	169.9		32	Ghana	137.5
8	Maldives	166.4			Italy	137.5
9	Lithuania	163.9		34	Panama	137.0
10	Turkmenistan	162.9		35	Georgia	136.4
11	South Africa	159.9		36	Ivory Coast	134.9
12	Russia	157.4			New Zealand	134.9
13	Cayman Islands	152.5		38	Poland	134.8
14	Mauritius	151.4		39	Malaysia	134.5
15	Botswana	150.0		40	Chile	134.4
16	Uruguay	149.9		41	Oman	133.4
17	Singapore	148.8		42	Bahrain	133.3
18	Vietnam	147.2		43	Mongolia	133.2
19	El Salvador	146.9		44	Slovakia	132.8
20	Estonia	145.4		45	Luxembourg	132.2
21	Sri Lanka	142.7		46	Argentina	132.1
22	Kazakhstan	142.3		47	Brunei	131.9
23	Qatar	141.9		48	Suriname	130.6
	Trinidad & Tobago	141.9		49	Colombia	129.9
25	Japan	141.4		50	South Korea	129.7

Landline telephones
Per 100 population, 2018

1	Monaco	112.1		23	Ireland	38.0
2	France	59.4		24	Canada	37.3
3	Malta	58.2			Iran	37.3
4	Hong Kong	56.9		26	New Zealand	37.1
5	Taiwan	55.5		27	Curaçao	36.1
6	Cayman Islands	54.9			Cyprus	36.1
7	Andorra	51.1		29	Belgium	35.8
	Germany	51.1		30	Singapore	34.8
9	South Korea	50.6		31	Bermuda	34.7
10	Japan	49.9		32	Netherlands	34.6
11	Portugal	49.5		33	Mauritius	34.3
12	Greece	48.3		34	Italy	33.6
13	Belarus	47.5			United States	33.6
	United Kingdom	47.5		36	Slovenia	33.4
15	Luxembourg	45.3			Uruguay	33.4
16	Barbados	44.7		38	Croatia	32.6
17	Austria	42.4		39	Australia	32.5
18	Spain	41.7			French Polynesia	32.5
19	Iceland	40.6		41	Hungary	31.1
20	Liechtenstein	40.2		42	Bahamas	29.4
21	Switzerland	38.7		43	Serbia	29.3
22	Israel	38.2		44	Montenegro	27.5

Broadband

Fixed-broadband subscribers per 100 population, 2018

1	Monaco	51.2	26	United States	33.8
2	Cayman Islands	49.3	27	Estonia	33.4
3	Switzerland	46.4	28	Japan	32.6
4	Andorra	46.3	29	Spain	32.5
5	France	44.8	30	Curaçao	31.9
6	Denmark	44.1	31	Hungary	31.7
	Liechtenstein	44.1	32	Finland	31.5
8	Malta	43.7	33	United Arab Emirates	31.4
9	Netherlands	43.4	34	Barbados	31.2
10	South Korea	41.6	35	Australia	30.7
11	Norway	41.3	36	Macau	30.6
12	Germany	41.1	37	Czech Republic	30.2
13	Iceland	40.6	38	Ireland	29.7
14	Sweden	39.9	39	Slovenia	29.5
15	United Kingdom	39.6	40	Israel	28.8
16	Belgium	39.2	41	China	28.5
17	Canada	39.0	42	Austria	28.4
18	Greece	37.7	43	Uruguay	28.3
19	Luxembourg	37.1	44	Lithuania	28.2
20	Portugal	36.9	45	Italy	28.1
21	Hong Kong	36.8	46	Singapore	28.0
22	Cyprus	36.3	47	Slovakia	27.7
23	Bermuda	36.2	48	Latvia	27.3
24	New Zealand	34.7	49	Croatia	27.1
25	Belarus	33.9	50	Bulgaria	27.0

Broadband speeds

Average download speed, Mbps, 2018

1	Singapore	60.4	23	Finland	24.0
2	Sweden	46.0		Germany	24.0
3	Denmark	44.0	25	New Zealand	23.8
4	Norway	40.1	26	Czech Republic	23.7
5	Romania	38.6	27	Slovenia	21.4
6	Belgium	36.7	28	Portugal	21.3
7	Netherlands	36.0	29	South Korea	20.6
8	Luxembourg	35.1	30	Bulgaria	20.2
9	Hungary	34.0	31	Poland	19.7
10	Switzerland	29.9	32	Canada	19.5
11	Japan	28.9	33	Iceland	18.9
12	Latvia	28.6	34	United Kingdom	18.6
13	Taiwan	28.1	35	Ireland	18.2
14	Estonia	27.9	36	Liechtenstein	17.7
15	Lithuania	27.2	37	Austria	17.5
	Spain	27.2	38	Barbados	17.1
17	Andorra	27.1		Thailand	17.1
18	Hong Kong	26.5	40	Macau	16.1
19	United States	25.9	41	Croatia	15.6
20	Slovakia	25.3	42	Italy	15.1
21	Madagascar	24.9	43	Moldova	13.9
22	France	24.2	44	Malta	13.6

Arts and entertainment

Music sales
Total including downloads, $bn, 2017

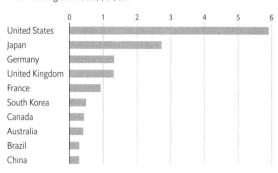

Revenue share, by type, %, 2017

■ Physical   ■ Digital   ■ Performance rights   ■ Synchronisation rights

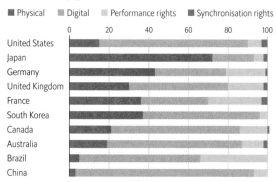

Book publishing
New titles per million population, 2018

1	Iceland	4,876	16	Serbia	1,350
2	Estonia	4,097	17	Norway	1,260
3	United Kingdom	2,800	18	Lithuania	1,098
4	Slovenia	2,357	19	Belarus	1,022
5	Italy	2,150	20	Moldova	973
6	Portugalᵃ	2,133	21	Germany	961
7	Denmark	2,082	22	Turkey	955
8	Belgium	1,955	23	Austria	952
9	Spain	1,851	24	Uruguay	937
10	Bulgaria	1,788	25	Saudi Arabiaᵃ	902
11	Finland	1,696	26	Russia	802
12	Franceᵇ	1,643	27	Greece	760
13	Czech Republic	1,540	28	Hong Kong	747
14	Latviaᵇ	1,509		Sweden	747
15	South Korea	1,421	30	Lebanon	668

a 2017   b 2016

Cinema attendances

Total visits, m, 2018

Visits per person, 2018

Cinema ticket prices

$, 2017 or latest

1	Bahrain	17.5		Hong Kong	10.7
2	Switzerland	15.4	17	United Kingdom	10.4
3	Macau	13.0	18	Austria	10.1
	Norway	13.0	19	Netherlands	9.4
5	Sweden	12.7	20	Belgium	9.3
6	Finland	12.6	21	Cyprus	9.1
7	Japan	11.7		Lebanon	9.1
8	United Arab Emirates	11.6	23	United States	9.0
9	Iceland	11.5	24	Ireland	8.6
10	Kuwait	11.4	25	New Zealand	8.5
11	Qatar	11.3	26	Canada	7.6
12	Australia	11.2	27	France	7.4
13	Israel	11.1	28	Greece	7.3
14	Denmark	10.9	29	Malta	7.2
15	Germany	10.7	30	Italy	7.1

The press

Press freedom[a]
Scores, 1=best, 100=worst, 2019

Most free		Least free	
1 Norway	7.8	**1** Turkmenistan	85.4
2 Finland	7.9	**2** North Korea	83.4
3 Sweden	8.3	**3** Eritrea	80.3
4 Netherlands	8.6	**4** China	78.9
5 Denmark	9.9	**5** Vietnam	74.9
6 Switzerland	10.5	**6** Sudan	72.5
7 New Zealand	10.8	**7** Syria	71.8
8 Jamaica	11.1	**8** Saudi Arabia	65.9
9 Belgium	12.1	**9** Laos	64.5
10 Costa Rica	12.2	**10** Iran	64.4
11 Estonia	12.3	**11** Cuba	63.8
12 Portugal	12.6	**12** Yemen	61.7
13 Germany	14.6	**13** Bahrain	61.3
14 Iceland	14.7	**14** Azerbaijan	59.1
15 Ireland	15.0	**15** Equatorial Guinea	58.4
16 Austria	15.3	**16** Somalia	57.2
17 Canada	15.7	**17** Egypt	56.5
Luxembourg	15.7	**18** Libya	55.8

Index of abuse against journalists[b]
2019, 100=worst

1 Syria	81.5	**14** Egypt	62.8
2 China	80.0	**15** Iran	59.6
3 Eritrea	72.2	**16** Colombia	59.0
4 Afghanistan	72.1	**17** Brazil	58.9
5 Yemen	69.4	**18** West Bank & Gaza	57.5
6 Mexico	68.5	**19** Pakistan	57.2
7 Azerbaijan	66.7	**20** Laos	56.4
8 Bahrain	65.6	**21** Central African Rep.	56.0
9 Saudi Arabia	65.5	Philippines	56.0
10 India	65.3	**23** Bangladesh	55.5
11 Vietnam	64.2	**24** Somalia	53.1
12 Turkey	63.9	**25** Ecuador	51.9
13 United States	63.0	North Korea	51.9

Number of journalists in prison
As of end 2018

1 Turkey	68	**10** Bahrain	5
2 China	47	Russia	5
3 Egypt	25	Syria	5
4 Saudi Arabia	18	**13** Morocco	4
5 Eritrea	16	Rwanda	4
6 Vietnam	11	**15** Israel[c]	3
7 Azerbaijan	10	Venezuela	3
8 Iran	9	**17** Congo-Kinshasa	2
9 Cameroon	8	Myanmar	2

a Based on 87 questions on topics such as media independence, censorship and transparency in 2018.   b Based on the intensity of abuse and violence against the media in 2018.   c Includes West Bank & Gaza.

Nobel prize winners: 1901–2019

Peace

1	United States	19
2	United Kingdom	12
3	France	9
4	Sweden	5
5	Belgium	4
	Germany	4

Medicine

1	United States	59
2	United Kingdom	26
3	Germany	15
4	France	8
5	Sweden	7

Literature

1	France	17
2	United States	13
3	United Kingdom	11
4	Germany	8
5	Sweden	7

Economics[a]

1	United States	42
2	United Kingdom	9
3	France	2
	Norway	2
	Sweden	2

Physics

1	United States	57
2	United Kingdom	21
3	Germany	19
4	France	11
5	Japan	7

Chemistry

1	United States	53
2	United Kingdom	26
3	Germany	16
4	France	8
5	Switzerland	7

Nobel prize winners: 1901–2019

By country of birth

1	United States	275			Egypt	6
2	United Kingdom	103			Israel	6
3	Germany	83	**26**	Finland	5
4	France	56			Ireland	5
5	Sweden	29			Ukraine	5
6	Japan	27	**29**	Argentina	4
7	Poland	26			Belarus	4
	Russia	26			Romania	4
9	Canada	19	**32**	Lithuania	3
	Italy	19			Mexico	3
	Switzerland	19			New Zealand	3
12	Austria	18			Pakistan	3
	Netherlands	18	**36**	Algeria	2
14	China	12			Bosnia & Herz.	2
	Norway	12			Chile	2
16	Denmark	11			Colombia	2
17	Australia	10			Guatemala	2
18	Belgium	9			Iran	2
	Hungary	9			Liberia	2
	India	9			Luxembourg	2
	South Africa	9			Portugal	2
22	Spain	7			South Korea	2
23	Czech Republic	6			Turkey	2

Notes: Prizes by country of residence at time awarded. When prizes have been shared in the same field, one credit given to each country.   a Since 1969.

Sports champions and cheats

World Cup winners and finalists
█ Winner █ Runner-up

Men's football (since 1930)

Brazil
Germany[a]
Italy
Argentina
France
Uruguay
England
Spain
Netherlands
Czechoslovakia[b]
Hungary
Croatia
Sweden

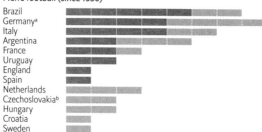

Women's football (since 1991)

United States
Germany
Japan
Norway
Brazil
China
Netherlands
Sweden

Men's cricket (Since 1975)

Australia
India
West Indies
England
Sri Lanka
Pakistan
New Zealand

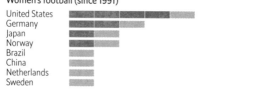

Women's cricket (Since 1973)

Australia
England
New Zealand
India
West Indies

Davis Cup, tennis (since 1900)[c]

United States
Australia
France
Great Britain
Sweden
Spain
Czech Republic[d]
Germany[a]
Russia
Croatia
Italy
Argentina
Serbia
Switzerland
South Africa

Note: Data as of May 2020.   a Including West Germany.   b Until 1993.
c Excludes finalists who have never won.   d Including Czechoslovakia.

Olympic games

Summer: 1896–2016

■ Gold   ■ Silver   ■ Bronze

	0	500	1,000	1,500	2,000	2,500
United States						
Soviet Union/ Unified team (1952–92)						
Great Britain						
China						
Germany						
France						
Italy						
Hungary						
East Germany (1968–88)						
Russia						
Australia						
Sweden						
Japan						
Finland						
South Korea						
Romania						
Netherlands						
Cuba						
Poland						
Canada						
West Germany (1968–88)						
Norway						
Bulgaria						
Switzerland						

Doping

Anti-doping rule violations, 2017

1	Italy	171		Kenya	29
2	France	128		Romania	29
3	United States	103	**22**	Colombia	26
4	Brazil	84		Portugal	26
5	Russia	82		Ukraine	26
6	China	62	**25**	Costa Rica	21
7	India	57	**26**	Netherlands	20
8	Belgium	54	**27**	Argentina	18
9	Spain	52	**28**	Bulgaria	16
10	Turkey	45		Chile	16
11	South Africa	43		Hungary	16
	United Kingdom	43		Thailand	16
13	Egypt	34	**32**	Czech Republic	15
	Germany	34	**33**	Slovakia	14
	Kazakhstan	34	**34**	Iran	13
16	Canada	33		Morocco	13
	South Korea	33		Norway	13
18	Azerbaijan	32		Peru	13
19	Australia	29			

Vices

Beer drinkers
Consumption, litres per person, 2018

1	Czech Republic	191.8	11	Slovenia	80.2
2	Austria	107.6	12	Netherlands	78.1
3	Germany	101.1	13	Latvia	76.8
4	Romania	98.9	14	Panama	76.7
5	Poland	98.2	15	Finland	76.6
6	Ireland	95.8	16	Lithuania	76.5
7	Spain	86.0	17	Bulgaria	76.4
8	Slovakia	83.5	18	Australia	76.3
9	Namibia	81.3	19	Hungary	74.8
10	Croatia	80.4	20	United States	73.5

Wine consumption

Litres, m, 2016			*Litres per person, 2016*		
1	United States	3,711	1	Andorra	50.5
2	France	2,710	2	Macedonia	41.8
3	Italy	2,240	3	France	41.7
4	Germany	2,010	4	Slovenia	39.7
5	China	1,750	5	Portugal	39.6
6	United Kingdom	1,290	6	Montenegro	37.0
7	Spain	1,030	7	Italy	36.9
8	Argentina	940	8	Albania	34.7
9	Russia	910	9	Switzerland	34.0
10	Australia	550	10	Serbia	31.8
11	Canada	506	11	Denmark	27.8
12	South Africa	420	12	Uruguay	27.5
13	Portugal	406	13	Bermuda	27.1
14	Romania	390	14	Austria	27.0
15	Japan	351	15	Belgium	26.1
16	Netherlands	316	16	Monaco	25.6
17	Brazil	310	17	Germany	24.2
18	Belgium	300	18	Croatia	23.8
19	Switzerland	290	19	Hungary	23.7
20	Serbia	280	20	Sweden	23.1
21	Austria	240	21	Australia	22.1
				Spain	22.1

Gambling losses
$ per adult, 2019

1	Australia	830	11	Canada	366
2	Hong Kong	828	12	Bermuda	365
3	Singapore	637	13	Denmark	339
4	Finland	485		Norway	339
5	Japan	473	15	Iceland	337
6	United States	463	16	United Kingdom	325
7	Ireland	420	17	Sweden	316
	New Zealand	420	18	Panama	310
9	Italy	383		Switzerland	310
10	Malta	373	20	Cyprus	300

Tourism

Most tourist arrivals
Number of arrivals, '000, 2018

#	Country	Value	#	Country	Value
1	France	89,322	21	Macau	18,493
2	Spain	82,773	22	Hungary	17,552
3	United States	79,746	23	India	17,423
4	China	62,900	24	Croatia	16,645
5	Italy	61,567	25	Portugal	16,186
6	Turkey	45,768	26	Indonesia	15,810
7	Mexico	41,313	27	Vietnam	15,498
8	Germany	38,881	28	South Korea	15,347
9	Thailand	38,178	29	Saudi Arabia	15,334
10	United Kingdom	36,316	30	Singapore	14,673
11	Japan	31,192	31	Ukraine	14,104
12	Austria	30,816	32	Denmark	12,749
13	Greece	30,123	33	Morocco	12,289
14	Hong Kong	29,263	34	Bahrain	12,045
15	Malaysia	25,832	35	Romania	11,720
16	Russia	24,551	36	Belarus	11,502
17	United Arab Emirates	21,286	37	Egypt	11,196
18	Canada	21,134	38	Ireland	10,926
19	Poland	19,622	39	Czech Republic	10,611
20	Netherlands	18,780	40	South Africa	10,472

Biggest tourist spenders
$bn, 2018

#	Country	Value	#	Country	Value
1	China	277.3	13	Hong Kong	26.5
2	United States	186.5	14	Netherlands	26.0
3	Germany	104.2	15	India	25.8
4	United Kingdom	68.9	16	Singapore	25.3
5	France	57.9	17	Brazil	22.2
6	Australia	42.4	18	Belgium	20.8
7	Russia	38.8	19	Switzerland	20.7
8	Italy	37.6	20	Norway	18.6
9	South Korea	34.8	21	Sweden	18.1
10	Canada	33.6	22	United Arab Emirates	18.0
11	Japan	28.1	23	Saudi Arabia	17.9
12	Spain	26.7	24	Thailand	14.7

Largest tourist receipts
$bn, 2018

#	Country	Value	#	Country	Value
1	United States	256.1	13	Turkey	37.1
2	Spain	81.3	14	India	29.1
3	France	73.1	15	Netherlands	25.9
4	Thailand	65.2	16	Austria	25.4
5	Germany	60.3	17	Portugal	24.1
6	Italy	51.6	18	Mexico	23.8
7	United Kingdom	48.5	19	Canada	22.0
8	Australia	47.3	20	Malaysia	21.8
9	Japan	45.3	21	Greece	21.6
10	Hong Kong	41.9	22	United Arab Emirates	21.4
11	China	40.4	23	Singapore	20.4
	Macau	40.4	24	Switzerland	20.3

Country profiles

ALGERIA

Area, sq km	2,381,741	Capital	Algiers
Arable as % of total land	3.1	Currency	Algerian dinar (AD)

People

Population, m	42.2	Life expectancy: men	76.3 yrs
Pop. per sq km	17.7	women	78.8 yrs
Average annual rate of change		Adult literacy	81.4
in pop. 2015–20, %	2.0	Fertility rate (per woman)	3.1
Pop. aged 0–19, 2020, %	37.4	Urban population, %	72.6
Pop. aged 65 and over, 2020, %	6.7		per 1,000 pop.
No. of men per 100 women	102.1	Crude birth rate	24.7
Human Development Index	75.9	Crude death rate	4.7

The economy

GDP	$174bn	GDP per head	$4,115
GDP	AD20,259bn	GDP per head in purchasing	
Av. ann. growth in real		power parity (USA=100)	24.7
GDP 2013–18	2.7%	Economic freedom index	46.9

Origins of GDP		**Components of GDP**	
	% of total		% of total
Agriculture	12.9	Private consumption	42.3
Industry, of which:	42.2	Public consumption	17.3
manufacturing	27.0	Investment	47.1
Services	44.5	Exports	25.6
		Imports	-32.3

Structure of employment

	% of total		% of labour force
Agriculture	9.9	Unemployed 2019	11.9
Industry	30.7	Av. ann. rate 2009–19	11.7
Services	59.4		

Energy

	m TOE		
Total output	171.0	Net energy imports as %	
Total consumption	60.5	of energy use	-183
Consumption per head			
kg oil equivalent	1,463		

Inflation and finance

			% change 2018–19
Consumer price			
inflation 2019	2.0%	Narrow money (M1)	-0.4
Av. ann. inflation 2014–19	4.6%	Broad money	-1.4
Deposit rate, Dec. 2019	1.75%		

Exchange rates

	end 2019		December 2019
			2010 = 100
AD per $	120.1	Effective rates	
AD per SDR	164.9	– nominal	77.6
AD per €	134.9	– real	95.5

Trade

Principal exports		Principal imports	
	$bn fob		*$bn cif*
Hydrocarbons	39.0	Capital goods	13.4
Semi-finished goods	2.2	Intermediate goods	11.0
Capital goods	0.1	Consumer goods	9.8
Raw materials	0.1	Food	8.6
Total incl. others	**41.8**	Total incl. others	**46.5**

Main export destinations		Main origins of imports	
	% of total		*% of total*
Italy	14.7	China	16.9
Spain	12.0	France	10.3
France	11.1	Italy	7.9
United States	9.2	South Africa	7.6

Balance of payments, reserves and debt, $bn

Visible exports fob	41.1	Change in reserves	-17.5
Visible imports fob	-48.6	Level of reserves	
Trade balance	-7.5	end Dec.	87.4
Invisibles inflows	4.6	No. months of import cover	15.9
Invisibles outflows	-17.2	Official gold holdings, m oz	5.6
Net transfers	3.4	Foreign debt	5.7
Current account balance	-16.7	– as % of GDP	3.3
– as % of GDP	-9.6	– as % of total exports	12.0
Capital balance	0.9	Debt service ratio	0.5
Overall balance	-15.7		

Health and education

Health spending, % of GDP	6.4	Education spending, % of GDP	...
Doctors per 1,000 pop.	1.7	Enrolment, %: primary	110
Hospital beds per 1,000 pop.	1.9	secondary	...
At least basic drinking water,		tertiary	51
% of pop.	93.6		

Society

No. of households, m	8.1	Cost of living, Dec. 2019	
Av. no. per household	5.2	New York = 100	42
Marriages per 1,000 pop.	...	Cars per 1,000 pop.	94
Divorces per 1,000 pop.	...	Telephone lines per 100 pop.	10.0
Religion, % of pop.		Mobile telephone subscribers	
Muslim	97.9	per 100 pop.	111.7
Non-religious	1.8	Internet access, %	49.0
Christian	0.2	Broadband subs per 100 pop.	7.3
Hindu	<0.1	Broadband speed, Mbps	1.2
Jewish	<0.1		
Other	<0.1		

ARGENTINA

Area, sq km	2,780,400	Capital	Buenos Aires
Arable as % of total land	14.3	Currency	Peso (P)

People

Population, m	44.4	Life expectancy: men	73.8 yrs
Pop. per sq km	16.0	women	80.4 yrs
Average annual rate of change		Adult literacy	99.0
in pop. 2015–20, %	1.0	Fertility rate (per woman)	2.3
Pop. aged 0–19, 2020, %	32.3	Urban population, %	91.9
Pop. aged 65 and over, 2020, %	11.4		per 1,000 pop.
No. of men per 100 women	95.3	Crude birth rate	17.1
Human Development Index	83.0	Crude death rate	7.6

The economy

GDP	$520bn	GDP per head	$11,684
GDP	P14,606bn	GDP per head in purchasing	
Av. ann. growth in real		power parity (USA=100)	32.8
GDP 2013–18	-0.4%	Economic freedom index	53.1

Origins of GDP		Components of GDP	
	% of total		% of total
Agriculture	7.2	Private consumption	65.1
Industry, of which:	27.2	Public consumption	16.1
manufacturing	15.0	Investment	21.0
Services	65.6	Exports	14.3
		Imports	-16.4

Structure of employment

	% of total		% of labour force
Agriculture	0.1	Unemployed 2019	9.2
Industry	21.4	Av. ann. rate 2009-19	9.8
Services	78.5		

Energy

	m TOE		
Total output	79.6	Net energy imports as %	
Total consumption	97.5	of energy use	18
Consumption per head			
kg oil equivalent	2,203		

Inflation and finance

Consumer price			% change 2018–19
inflation 2019	53.5%	Narrow money (M1)	28.5
Av. ann. inflation 2014–19	37.4%	Broad money	37.2
Deposit rate, Dec. 2019	46.85%		

Exchange rates

	end 2019		December 2019
P per $	59.8	Effective rates	2010 = 100
P per SDR	82.7	– nominal	...
P per €	67.2	– real	...

Trade

Principal exports		Principal imports	
	$bn fob		*$bn cif*
Processed agricultural products	22.9	Intermediate goods	20.4
Manufactures	20.5	Capital goods	12.2
Primary products	14.0	Consumer goods	8.5
Fuels & energy	4.2	Fuels	6.5
Total incl. others	**61.8**	Total incl. others	**65.5**

Main export destinations		Main origins of imports	
	% of total		*% of total*
Brazil	18.3	Brazil	24.0
China	6.8	China	18.4
United States	6.8	United States	11.7
Chile	5.0	Germany	5.1

Balance of payments, reserves and debt, $bn

Visible exports fob	61.8	Change in reserves	10.9
Visible imports fob	-62.5	Level of reserves	
Trade balance	-0.7	end Dec.	66.2
Invisibles inflows	21.3	No. months of import cover	7.1
Invisibles outflows	-49.1	Official gold holdings, m oz	1.8
Net transfers	1.3	Foreign debt	280.5
Current account balance	-27.3	– as % of GDP	54.0
– as % of GDP	-5.2	– as % of total exports	335.9
Capital balance	11.1	Debt service ratio	44.8
Overall balance	-17.1		

Health and education

Health spending, % of GDP	9.1	Education spending, % of GDP	5.5
Doctors per 1,000 pop.	4.0	Enrolment, %: primary	110
Hospital beds per 1,000 pop.	5.0	secondary	109
At least basic drinking water,		tertiary	90
% of pop.	99.1		

Society

No. of households, m	14.3	Cost of living, Dec. 2019	
Av. no. per household	3.1	New York = 100	35
Marriages per 1,000 pop.	2.9	Cars per 1,000 pop.	256
Divorces per 1,000 pop.	...	Telephone lines per 100 pop.	22.0
Religion, % of pop.		Mobile telephone subscribers	
Christian	85.2	per 100 pop.	132.1
Non-religious	12.2	Internet access, %	74.3
Other	1.1	Broadband subs per 100 pop.	19.1
Muslim	1.0	Broadband speed, Mbps	3.2
Jewish	0.5		
Hindu	<0.1		

AUSTRALIA

Area, sq km	7,741,220	Capital	Canberra
Arable as % of total land	6.0	Currency	Australian dollar (A$)

People

Population, m	24.9	Life expectancy: men	82.1 yrs
Pop. per sq km	3.2	women	85.8 yrs
Average annual rate of change		Adult literacy	...
in pop. 2015–20, %	1.3	Fertility rate (per woman)	1.8
Pop. aged 0–19, 2020, %	25.3	Urban population, %	86.0
Pop. aged 65 and over, 2020, %	16.2		per 1,000 pop.
No. of men per 100 women	99.2	Crude birth rate	12.9
Human Development Index	93.8	Crude death rate	6.7

The economy

GDP	$1,434bn	GDP per head	$57,374
GDP	A$1,898bn	GDP per head in purchasing	
Av. ann. growth in real		power parity (USA=100)	82.3
GDP 2013–18	2.6%	Economic freedom index	82.6

Origins of GDP		**Components of GDP**	
	% of total		% of total
Agriculture	2.6	Private consumption	56.4
Industry, of which:	25.9	Public consumption	18.7
manufacturing	6.2	Investment	24.5
Services	71.7	Exports	21.8
		Imports	-21.4

Structure of employment

	% of total		% of labour force
Agriculture	2.6	Unemployed 2019	5.3
Industry	19.8	Av. ann. rate 2009-19	5.3
Services	77.6		

Energy

	m TOE		
Total output	417.3	Net energy imports as %	
Total consumption	153.4	of energy use	-172
Consumption per head			
kg oil equivalent	6,275		

Inflation and finance

			% change 2018–19
Consumer price			
inflation 2019	1.6%	Narrow money (M1)	3.6
Av. ann. inflation 2014–19	1.7%	Broad money	4.1
Deposit rate, Dec. 2019	1.15%		

Exchange rates

	end 2019		December 2019
A$ per $	1.43	Effective rates	2010 = 100
A$ per SDR	1.97	– nominal	85.7
A$ per €	1.61	– real	85.2

Trade

Principal exports		Principal imports	
	$bn fob		*$bn cif*
Crude materials	91.1	Machinery & transport equip.	90.6
Fuels	79.7	Mineral fuels	33.1
Food	26.4	Miscellaneous manuf. articles	24.3
Machinery & transport equip.	14.3	Manufactured goods	22.3
Total incl. others	**257.1**	Total incl. others	**227.0**

Main export destinations		Main origins of imports	
	% of total		*% of total*
China	33.8	China	25.9
Japan	16.0	United States	11.0
South Korea	6.5	Japan	7.9
India	4.6	Germany	5.3

Balance of payments, reserves and aid, $bn

Visible exports fob	257.8	Overall balance	-11.1
Visible imports fob	-236.9	Change in reserves	-12.7
Trade balance	20.9	Level of reserves	
Invisibles inflows	118.1	end Dec.	53.9
Invisibles outflows	-167.6	No. months of import cover	1.6
Net transfers	-0.7	Official gold holdings, m oz	2.2
Current account balance	-29.2	Aid given	3.1
– as % of GDP	-2.0	– as % of GNI	0.2
Capital balance	23.9		

Health and education

Health spending, % of GDP	9.2	Education spending, % of GDP	5.3
Doctors per 1,000 pop.	3.7	Enrolment, %: primary	100
Hospital beds per 1,000 pop.	3.8	secondary	150
At least basic drinking water,		tertiary	113
% of pop.	100		

Society

No. of households, m	9.2	Cost of living, Dec. 2019	
Av. no. per household	2.7	New York = 100	82
Marriages per 1,000 pop.	4.6	Cars per 1,000 pop.	560
Divorces per 1,000 pop.	2.0	Telephone lines per 100 pop.	32.5
Religion, % of pop.		Mobile telephone subscribers	
Christian	67.3	per 100 pop.	113.6
Non-religious	24.2	Internet access, %	86.6
Other	4.2	Broadband subs per 100 pop.	30.7
Muslim	2.4	Broadband speed, Mbps	11.7
Hindu	1.4		
Jewish	0.5		

AUSTRIA

Area, sq km	83,879	Capital	Vienna
Arable as % of total land	16.3	Currency	Euro (€)

People

Population, m	8.9	Life expectancy: men	79.9 yrs
Pop. per sq km	106.1	women	84.2 yrs
Average annual rate of change		Adult literacy	...
in pop. 2015–20, %	0.7	Fertility rate (per woman)	1.5
Pop. aged 0–19, 2020, %	19.4	Urban population, %	58.3
Pop. aged 65 and over, 2020, %	19.2		per 1,000 pop.
No. of men per 100 women	97.2	Crude birth rate	9.9
Human Development Index	91.4	Crude death rate	10.0

The economy

GDP	$455bn	GDP per head	$51,462
GDP	€386bn	GDP per head in purchasing	
Av. ann. growth in real		power parity (USA=100)	88.3
GDP 2013–18	1.7%	Economic freedom index	73.3

Origins of GDP		**Components of GDP**	
	% of total		% of total
Agriculture	1.3	Private consumption	51.8
Industry, of which:	28.8	Public consumption	19.3
manufacturing	18.9	Investment	25.1
Services	69.9	Exports	55.8
		Imports	-52.0

Structure of employment

	% of total		% of labour force
Agriculture	3.6	Unemployed 2019	4.8
Industry	25.3	Av. ann. rate 2009–19	4.7
Services	71.2		

Energy

	m TOE		
Total output	13.4	Net energy imports as %	
Total consumption	37.7	of energy use	64
Consumption per head			
kg oil equivalent	4,312		

Inflation and finance

			% change 2018–19
Consumer price			
inflation 2019	1.5%	Narrow money (M1)	1.8
Av. ann. inflation 2014–19	1.5%	Broad money	5.2
Deposit rate, Dec. 2019	0.14%		

Exchange rates

	end 2019		December 2019
€ per $	0.89	Effective rates	2010 = 100
€ per SDR	1.23	– nominal	100.6
		– real	101.4

Trade

Principal exports

	$bn fob
Machinery & transport equip.	75.6
Chemicals & related products	23.9
Food, drink & tobacco	13.4
Raw materials	6.2
Total incl. others	**184.8**

Principal imports

	$bn cif
Machinery & transport equip.	71.0
Chemicals & related products	25.5
Mineral fuels & lubricants	15.1
Food, drink & tobacco	13.5
Total incl. others	**193.8**

Main export destinations

	% of total
Germany	29.5
Italy	6.4
United States	6.3
Switzerland	4.9
EU28	71.4

Main origins of imports

	% of total
Germany	41.0
Italy	6.0
Switzerland	4.7
Czech Republic	4.5
EU28	77.6

Balance of payments, reserves and aid, $bn

Visible exports fob	179.1	Overall balance	2.6
Visible imports fob	-174.7	Change in reserves	1.7
Trade balance	4.4	Level of reserves	
Invisibles inflows	108.6	end Dec.	23.2
Invisibles outflows	-97.6	No. months of import cover	1.0
Net transfers	-4.6	Official gold holdings, m oz	9.0
Current account balance	10.8	Aid given	1.2
– as % of GDP	2.4	– as % of GNI	0.3
Capital balance	-10.1		

Health and education

Health spending, % of GDP	10.4	Education spending, % of GDP	5.5
Doctors per 1,000 pop.	5.2	Enrolment, %: primary	103
Hospital beds per 1,000 pop.	7.6	secondary	100
At least basic drinking water,		tertiary	85
% of pop.	100		

Society

No. of households, m	3.9	Cost of living, Dec. 2019	
Av. no. per household	2.3	New York = 100	83
Marriages per 1,000 pop.	5.3	Cars per 1,000 pop.	546
Divorces per 1,000 pop.	1.8	Telephone lines per 100 pop.	42.4
Religion, % of pop.		Mobile telephone subscribers	
Christian	80.4	per 100 pop.	123.5
Non-religious	13.5	Internet access, %	87.5
Muslim	5.4	Broadband subs per 100 pop.	28.4
Other	0.5	Broadband speed, Mbps	17.5
Jewish	0.2		
Hindu	<0.1		

BANGLADESH

Area, sq km	147,630	Capital	Dhaka
Arable as % of total land	59.6	Currency	Taka (Tk)

People

Population, m	161.4	Life expectancy: men	71.8 yrs
Pop. per sq km	1,093.7	women	75.6 yrs
Average annual rate of change		Adult literacy	73.9
in pop. 2015–20, %	1.1	Fertility rate (per woman)	2.1
Pop. aged 0–19, 2020, %	36.2	Urban population, %	36.6
Pop. aged 65 and over, 2020, %	5.2		per 1,000 pop.
No. of men per 100 women	102.2	Crude birth rate	18.4
Human Development Index	61.4	Crude death rate	5.5

The economy

GDP	$274bn	GDP per head	$1,698
GDP	Tk24,095bn	GDP per head in purchasing	
Av. ann. growth in real		power parity (USA=100)	7.0
GDP 2013–18	7.0%	Economic freedom index	56.4

Origins of GDP		**Components of GDP**	
	% of total		% of total
Agriculture	13.8	Private consumption	70.8
Industry, of which:	30.4	Public consumption	6.4
manufacturing	19.0	Investment	31.2
Services	55.9	Exports	14.8
		Imports	-23.4

Structure of employment

	% of total		% of labour force
Agriculture	38.6	Unemployed 2019	4.3
Industry	21.3	Av. ann. rate 2009–19	4.2
Services	40.2		

Energy

	m TOE		
Total output	28.5	Net energy imports as %	
Total consumption	35.8	of energy use	20
Consumption per head			
kg oil equivalent	217		

Inflation and finance

			% change 2018–19
Consumer price			
inflation 2019	5.7%	Narrow money (M1)	7.0
Av. ann. inflation 2014–19	5.8%	Broad money	11.9
Deposit rate, Dec. 2019	7.09%		

Exchange rates

	end 2019		December 2019
Tk per $	84.90	Effective rates	2010 = 100
Tk per SDR	117.40	– nominal	...
Tk per €	95.39	– real	...

Trade

Principal exports		**Principal imports**	
	$bn fob		*$bn cif*
Clothing	23.7	Textiles & yarns	5.9
Jute goods	0.7	Capital machinery	5.1
Fish & fish products	0.4	Iron & steel	4.9
Leather	0.1	Fuels	4.0
Total incl. others	**33.6**	Total incl. others	**55.4**

Main export destinations		**Main origins of imports**	
	% of total		*% of total*
Germany	13.1	China	23.7
United States	12.4	India	15.6
United Kingdom	8.5	Singapore	5.2
Spain	5.7	Indonesia	3.5

Balance of payments, reserves and debt, $bn

Visible exports fob	38.7	Change in reserves	-1.4
Visible imports fob	-55.6	Level of reserves	
Trade balance	-16.9	end Dec.	32.0
Invisibles inflows	5.6	No. months of import cover	5.7
Invisibles outflows	-12.4	Official gold holdings, m oz	0.4
Net transfers	16.1	Foreign debt	52.1
Current account balance	-7.6	– as % of GDP	19.0
– as % of GDP	-2.8	– as % of total exports	87.1
Capital balance	7.9	Debt service ratio	4.7
Overall balance	-1.1		

Health and education

Health spending, % of GDP	2.3	Education spending, % of GDP	2.0
Doctors per 1,000 pop.	0.3	Enrolment, %: primary	116
Hospital beds per 1,000 pop.	0.8	secondary	73
At least basic drinking water,		tertiary	21
% of pop.	97.0		

Society

No. of households, m	33.3	Cost of living, Dec. 2019	
Av. no. per household	4.8	New York = 100	66
Marriages per 1,000 pop.	...	Cars per 1,000 pop.	3
Divorces per 1,000 pop.	...	Telephone lines per 100 pop.	0.9
Religion, % of pop. of pop.		Mobile telephone subscribers	
Muslim	89.8	per 100 pop.	100.2
Hindu	9.1	Internet access, %	15.0
Other	0.9	Broadband subs per 100 pop.	6.3
Christian	0.2	Broadband speed, Mbps	2.0
Jewish	<0.1		
Non-religious	<0.1		

BELGIUM

Area, sq km	30,530	Capital	Brussels
Arable as % of total land	28.1	Currency	Euro (€)

People

Population, m	11.5	Life expectancy: men	80.0 yrs
Pop. per sq km	376.7	women	84.3 yrs
Average annual rate of change		Adult literacy	...
in pop. 2015–20, %	0.5	Fertility rate (per woman)	1.7
Pop. aged 0–19, 2020, %	22.6	Urban population, %	98.0
Pop. aged 65 and over, 2020, %	19.3		per 1,000 pop.
No. of men per 100 women	98.3	Crude birth rate	10.9
Human Development Index	91.9	Crude death rate	9.8

The economy

GDP	$543bn	GDP per head	$47,519
GDP	€451bn	GDP per head in purchasing	
Av. ann. growth in real		power parity (USA=100)	81.9
GDP 2013–18	1.7%	Economic freedom index	68.9

Origins of GDP		**Components of GDP**	
	% of total		% of total
Agriculture	0.6	Private consumption	51.6
Industry, of which:	21.4	Public consumption	23.1
manufacturing	13.8	Investment	25.5
Services	78.0	Exports	82.6
		Imports	-82.7

Structure of employment

	% of total		% of labour force
Agriculture	1.0	Unemployed 2019	5.9
Industry	20.8	Av. ann. rate 2009–19	5.6
Services	78.2		

Energy

	m TOE		
Total output	14.5	Net energy imports as %	
Total consumption	67.0	of energy use	78
Consumption per head			
kg oil equivalent	5,864		

Inflation and finance

		% change 2018–19	
Consumer price			
inflation 2019	1.2%	Narrow money (M1)	1.8
Av. ann. inflation 2014–19	1.6%	Broad money	5.2
Deposit rate, Dec. 2019	0.11%		

Exchange rates

	end 2019		December 2019
€ per $	0.89	Effective rates	2010 = 100
€ per SDR	1.23	– nominal	101.6
		– real	98.9

Trade

Principal exports		Principal imports	
	$bn fob		*$bn cif*
Chemicals & related products	142.3	Chemicals & related products	114.7
Machinery & transport equip.	102.5	Machinery & transport equip.	112.5
Mineral fuels & lubricants	44.2	Mineral fuels & lubricants	62.9
Food, drink & tobacco	43.3	Food, drink & tobacco	36.6
Total incl. others	**468.6**	Total incl. others	**454.7**

Main export destinations		Main origins of imports	
	% of total		*% of total*
Germany	17.7	Netherlands	17.7
France	14.4	Germany	12.9
Netherlands	12.2	France	9.3
United Kingdom	7.9	United States	6.8
EU28	73.0	EU28	64.6

Balance of payments, reserves and debt, $bn

Visible exports fob	324.9	Overall balance	0.9
Visible imports fob	-325.7	Change in reserves	0.7
Trade balance	-0.8	Level of reserves	
Invisibles inflows	194.2	end Dec.	26.8
Invisibles outflows	-192.8	No. months of import cover	0.6
Net transfers	-8.0	Official gold holdings, m oz	7.3
Current account balance	-7.4	Aid given	2.3
– as % of GDP	-1.4	– as % of GNI	0.4
Capital balance	10.0		

Health and education

Health spending, % of GDP	10.3	Education spending, % of GDP	6.5
Doctors per 1,000 pop.	3.1	Enrolment, %: primary	104
Hospital beds per 1,000 pop.	6.2	secondary	159
At least basic drinking water,		tertiary	80
% of pop.	100		

Society

No. of households, m	4.8	Cost of living, Dec. 2019	
Av. no. per household	2.4	New York = 100	73
Marriages per 1,000 pop.	3.9	Cars per 1,000 pop.	497
Divorces per 1,000 pop.	2.0	Telephone lines per 100 pop.	35.8
Religion, % of pop.		Mobile telephone subscribers	
Christian	64.2	per 100 pop.	99.7
Non-religious	29.0	Internet access, %	88.7
Muslim	5.9	Broadband subs per 100 pop.	39.2
Other	0.6	Broadband speed, Mbps	36.7
Jewish	0.3		
Hindu	<0.1		

BRAZIL

Area, sq km	8,515,770	Capital	Brasília
Arable as % of total land	9.7	Currency	Real (R)

People

Population, m	209.5	Life expectancy: men	73.0 yrs
Pop. per sq km	24.6	women	80.1 yrs
Average annual rate of change		Adult literacy	93.2
in pop. 2015–20, %	0.8	Fertility rate (per woman)	1.7
Pop. aged 0–19, 2020, %	28.3	Urban population, %	86.6
Pop. aged 65 and over, 2020, %	9.6		per 1,000 pop.
No. of men per 100 women	96.6	Crude birth rate	14.1
Human Development Index	76.1	Crude death rate	6.8

The economy

GDP	$1,869bn	GDP per head	$8,921
GDP	R6,825bn	GDP per head in purchasing	
Av. ann. growth in real		power parity (USA=100)	25.6
GDP 2013–18	-0.9%	Economic freedom index	53.7

Origins of GDP		**Components of GDP**	
	% of total		% of total
Agriculture	5.1	Private consumption	64.3
Industry, of which:	21.6	Public consumption	19.7
manufacturing	11.3	Investment	15.4
Services	73.3	Exports	14.8
		Imports	-14.3

Structure of employment

	% of total		% of labour force
Agriculture	9.2	Unemployed 2019	12.3
Industry	19.8	Av. ann. rate 2009–19	12.1
Services	71.0		

Energy

	m TOE		
Total output	283.0	Net energy imports as %	
Total consumption	317.6	of energy use	11
Consumption per head			
kg oil equivalent	1,517		

Inflation and finance

			% change 2018–19
Consumer price			
inflation 2019	3.7%	Narrow money (M1)	3.3
Av. ann. inflation 2014–19	5.7%	Broad money	9.1
Deposit rate, Dec. 2019	4.10%		

Exchange rates

	end 2019		December 2019
			2010 = 100
R per $	4.03	Effective rates	
R per SDR	5.57	– nominal	63.1
R per €	4.53	– real	67.9

Trade

Principal exports		Principal imports	
	$bn fob		*$bn cif*
Primary products	119.2	Intermediate products & raw	
Manufactured goods	86.1	materials	105.0
Semi-manufactured goods	30.5	Capital goods	28.6
		Consumer goods	25.6
		Fuels & lubricants	22.0
Total incl. others	**239.3**	Total incl. others	**181.2**

Main export destinations		Main origins of imports	
	% of total		*% of total*
China	26.7	China	20.3
United States	12.2	United States	17.2
Argentina	6.2	Argentina	6.5
Netherlands	5.5	Germany	6.2

Balance of payments, reserves and debt, $bn

Visible exports fob	239.5	Change in reserves	0.8
Visible imports fob	-186.5	Level of reserves	
Trade balance	53.0	end Dec.	374.7
Invisibles inflows	48.7	No. months of import cover	13.6
Invisibles outflows	-143.2	Official gold holdings, m oz	2.2
Net transfers	0.0	Foreign debt	557.8
Current account balance	-41.5	– as % of GDP	29.6
– as % of GDP	-2.2	– as % of total exports	191.6
Capital balance	45.8	Debt service ratio	32.8
Overall balance	2.9		

Health and education

Health spending, % of GDP	9.5	Education spending, % of GDP	6.2
Doctors per 1,000 pop.	2.2	Enrolment, %: primary	115
Hospital beds per 1,000 pop.	2.2	secondary	101
At least basic drinking water,		tertiary	51
% of pop.	98.2		

Society

No. of households, m	64.8	Cost of living, Dec. 2019	
Av. no. per household	3.2	New York = 100	54
Marriages per 1,000 pop.	...	Cars per 1,000 pop.	184
Divorces per 1,000 pop.	...	Telephone lines per 100 pop.	18.3
Religion, % of pop.		Mobile telephone subscribers	
Christian	88.9	per 100 pop.	98.8
Non-religious	7.9	Internet access, %	70.4
Other	3.1	Broadband subs per 100 pop.	14.9
Hindu	<0.1	Broadband speed, Mbps	2.6
Jewish	<0.1		
Muslim	<0.1		

BULGARIA

Area, sq km	111,000	Capital	Sofia
Arable as % of total land	32.2	Currency	Lev (BGL)

People

Population, m	7.1	Life expectancy: men	72.1 yrs
Pop. per sq km	64.0	women	79.1 yrs
Average annual rate of change		Adult literacy	98.4
in pop. 2015–20, %	-0.7	Fertility rate (per woman)	1.6
Pop. aged 0–19, 2020, %	19.2	Urban population, %	75.0
Pop. aged 65 and over, 2020, %	21.5		per 1,000 pop.
No. of men per 100 women	94.4	Crude birth rate	9.0
Human Development Index	81.6	Crude death rate	15.6

The economy

GDP	$65bn	GDP per head	$9,273
GDP	BGL108bn	GDP per head in purchasing	
Av. ann. growth in real		power parity (USA=100)	35.0
GDP 2013–18	3.2%	Economic freedom index	70.2

Origins of GDP		**Components of GDP**	
	% of total		% of total
Agriculture	3.9	Private consumption	60.7
Industry, of which:	25.8	Public consumption	16.7
manufacturing	15.5	Investment	21.6
Services	70.3	Exports	66.9
		Imports	-64.3

Structure of employment

	% of total		% of labour force
Agriculture	6.4	Unemployed 2019	5.2
Industry	30.1	Av. ann. rate 2009-19	4.3
Services	63.5		

Energy

	m TOE		
Total output	11.5	Net energy imports as %	
Total consumption	19.5	of energy use	41
Consumption per head			
kg oil equivalent	2,753		

Inflation and finance

Consumer price			% change 2018–19
inflation 2019	2.5%	Narrow money (M1)	-1.4
Av. ann. inflation 2014–19	0.8%	Broad money	9.9
Deposit rate, Dec. 2019	0.02%		

Exchange rates

	end 2019		December 2019
			2010 = 100
BGL per $	1.74	Effective rates	
BGL per SDR	2.41	– nominal	109.6
BGL per €	1.96	– real	102.3

Trade

Principal exports	$bn fob	Principal imports	$bn cif
Raw materials	13.5	Raw materials	14.0
Capital goods	8.5	Capital goods	10.1
Consumer goods	8.4	Consumer goods	8.4
Mineral fuels & lubricants	2.5	Mineral fuels & lubricants	4.5
Total incl. others	**33.8**	Total incl. others	**38.0**

Main export destinations	% of total	Main origins of imports	% of total
Germany	14.6	Germany	12.4
Italy	8.5	Russia	9.7
Romania	8.4	Italy	7.5
Turkey	7.7	Romania	6.9
EU28	67.6	EU28	63.6

Balance of payments, reserves and debt, $bn

Visible exports fob	32.7	Change in reserves	0.3
Visible imports fob	-34.9	Level of reserves	
Trade balance	-2.2	end Dec.	28.7
Invisibles inflows	12.3	No. months of import cover	7.9
Invisibles outflows	-8.9	Official gold holdings, m oz	1.3
Net transfers	2.3	Foreign debt	39.9
Current account balance	3.5	– as % of GDP	60.2
– as % of GDP	5.4	– as % of total exports	83.8
Capital balance	-1.6	Debt service ratio	15.1
Overall balance	1.5		

Health and education

Health spending, % of GDP	8.1	Education spending, % of GDP	...
Doctors per 1,000 pop.	4.0	Enrolment, %: primary	89
Hospital beds per 1,000 pop.	6.8	secondary	98
At least basic drinking water,		tertiary	71
% of pop.	99.1		

Society

No. of households, m	2.7	Cost of living, Dec. 2019	
Av. no. per household	2.6	New York = 100	52
Marriages per 1,000 pop.	4.1	Cars per 1,000 pop.	466
Divorces per 1,000 pop.	1.5	Telephone lines per 100 pop.	15.9
Religion, % of pop.		Mobile telephone subscribers	
Christian	82.1	per 100 pop.	118.9
Muslim	13.7	Internet access, %	64.8
Non-religious	4.2	Broadband subs per 100 pop.	27.0
Hindu	<0.1	Broadband speed, Mbps	20.2
Jewish	<0.1		
Other	<0.1		

CAMEROON

Area, sq km	475,440	Capital	Yaoundé
Arable as % of total land	13.1	Currency	CFA franc (CFAfr)

People

Population, m	25.2	Life expectancy: men	59.0 yrs
Pop. per sq km	53.0	women	61.7 yrs
Average annual rate of change		Adult literacy	77.1
in pop. 2015–20, %	2.6	Fertility rate (per woman)	4.6
Pop. aged 0–19, 2020, %	52.8	Urban population, %	56.4
Pop. aged 65 and over, 2020, %	2.7		per 1,000 pop.
No. of men per 100 women	100.1	Crude birth rate	35.6
Human Development Index	56.3	Crude death rate	8.5

The economy

GDP	$39bn	GDP per head	$1,534
GDP	CFAfr21,493bn	GDP per head in purchasing	
Av. ann. growth in real		power parity (USA=100)	6.0
GDP 2013–18	4.8%	Economic freedom index	53.6

Origins of GDP		**Components of GDP**	
	% of total		% of total
Agriculture	15.7	Private consumption	70.5
Industry, of which:	28.1	Public consumption	11.1
manufacturing	15.5	Investment	22.8
Services	56.2	Exports	19.3
		Imports	-23.7

Structure of employment

	% of total		% of labour force
Agriculture	43.4	Unemployed 2019	3.4
Industry	15.0	Av. ann. rate 2009-19	3.4
Services	41.6		

Energy

	m TOE		
Total output	6.1	Net energy imports as %	
Total consumption	3.7	of energy use	-62
Consumption per head			
kg oil equivalent	156		

Inflation and finance

		% change 2018–19	
Consumer price			
inflation 2019	2.5%	Narrow money (M1)	13.0
Av. ann. inflation 2014–19	1.5%	Broad money	11.1
Deposit rate, Oct. 2018	2.45%		

Exchange rates

	end 2019		December 2019
CFAfr per $	583.9	Effective rates	2010 = 100
CFAfr per SDR	807.4	– nominal	102.0
CFAfr per €	656.1	– real	96.3

Trade

Principal exports	
	$bn fob
Fuels	1.6
Timber	0.6
Cocoa beans & products	0.4
Cotton	0.2
Total incl. others	**3.8**

Principal imports	
	$bn cif
Food, drink & tobacco	1.3
Fuels	1.3
Machinery & transport equip.	1.3
Chemicals	0.8
Total incl. others	**6.1**

Main export destinations	
	% of total
China	19.8
Italy	16.5
Netherlands	11.1
India	8.4

Main origins of imports	
	% of total
China	17.8
France	8.2
Nigeria	4.6
Italy	3.9

Balance of payments, reserves and debt, $bn

Visible exports fob	5.2	Change in reserves	0.3
Visible imports fob	-5.7	Level of reserves	
Trade balance	-0.5	end Dec.	3.5
Invisibles inflows	2.3	No. months of import cover	4.4
Invisibles outflows	-3.7	Official gold holdings, m oz	0.0
Net transfers	0.5	Foreign debt	10.7
Current account balance	-1.4	– as % of GDP	27.7
– as % of GDP	-3.6	– as % of total exports	136.2
Capital balance	1.8	Debt service ratio	13.7
Overall balance	0.3		

Health and education

Health spending, % of GDP	4.7	Education spending, % of GDP	3.1
Doctors per 1,000 pop.	...	Enrolment, %: primary	103
Hospital beds per 1,000 pop.	1.3	secondary	60
At least basic drinking water,		tertiary	13
% of pop.	60.4		

Society

No. of households, m	5.0	Cost of living, Dec. 2019	
Av. no. per household	5.0	New York = 100	...
Marriages per 1,000 pop.	...	Cars per 1,000 pop.	11
Divorces per 1,000 pop.	...	Telephone lines per 100 pop.	3.6
Religion, % of pop.		Mobile telephone subscribers	
Christian	70.3	per 100 pop.	73.2
Muslim	18.3	Internet access, %	23.2
Other	6.0	Broadband subs per 100 pop.	0.1
Non-religious	5.3	Broadband speed, Mbps	1.3
Hindu	<0.1		
Jewish	<0.1		

CANADA

Area, sq km[a]	9,984,670	Capital	Ottawa
Arable as % of total land	4.8	Currency	Canadian dollar (C$)

People

Population, m	37.1	Life expectancy: men	81.2 yrs
Pop. per sq km	3.7	women	84.7 yrs
Average annual rate of change		Adult literacy	...
in pop. 2015–20, %	0.9	Fertility rate (per woman)	1.5
Pop. aged 0–19, 2020, %	21.0	Urban population, %	81.4
Pop. aged 65 and over, 2020, %	18.1		per 1,000 pop.
No. of men per 100 women	98.5	Crude birth rate	10.5
Human Development Index	92.2	Crude death rate	7.9

The economy

GDP	$1,713bn	GDP per head	$46,233
GDP	C$2,219bn	GDP per head in purchasing	
Av. ann. growth in real		power parity (USA=100)	76.6
GDP 2013–18	1.9%	Economic freedom index	78.2

Origins of GDP		**Components of GDP**	
	% of total		% of total
Agriculture	2.1	Private consumption	58.0
Industry, of which:	27.9	Public consumption	21.0
manufacturing	10.4	Investment	23.1
Services	70.2	Exports	32.1
		Imports	-34.1

Structure of employment

	% of total		% of labour force
Agriculture	1.5	Unemployed 2019	5.8
Industry	19.5	Av. ann. rate 2009–19	5.6
Services	79.1		

Energy

	m TOE		
Total output	552.0	Net energy imports as %	
Total consumption	379.7	of energy use	-45
Consumption per head			
kg oil equivalent	10,366		

Inflation and finance

			% change 2018–19
Consumer price			
inflation 2019	1.9%	Narrow money (M1)	3.2
Av. ann. inflation 2014–19	1.7%	Broad money	7.0
Central bank policy rate, Dec. 2019	1.75%		

Exchange rates

	end 2019		December 2019
			2010 = 100
C$ per $	1.31	Effective rates	
C$ per SDR	1.81	– nominal	84.1
C$ per €	1.47	– real	82.0

Trade

Principal exports		**Principal imports**	
	$bn fob		*$bn cif*
Energy products	85.7	Consumer goods	93.6
Motor vehicles & parts	69.8	Motor vehicles & parts	87.8
Consumer goods	51.4	Electronic & electrical equip.	55.0
Metal & mineral products	49.9	Industrial machinery & equip.	52.6
Total incl. others	**451.5**	Total incl. others	**468.6**

Main export destinations		**Main origins of imports**	
	% of total		*% of total*
United States	75.5	United States	51.3
China	4.7	China	12.7
United Kingdom	2.8	Mexico	6.2
Japan	2.2	Germany	3.2
EU28	7.4	EU28	12.3

Balance of payments, reserves and aid, $bn

Visible exports fob	451.6	Overall balance	-1.7
Visible imports fob	-468.7	Change in reserves	-2.8
Trade balance	-17.1	Level of reserves	
Invisibles inflows	201.3	end Dec.	83.9
Invisibles outflows	-224.6	No. months of import cover	1.5
Net transfers	-2.6	Official gold holdings, m oz	0.0
Current account balance	-43.0	Aid given	4.6
– as % of GDP	-2.5	– as % of GNI	0.3
Capital balance	33.8		

Health and education

Health spending, % of GDP	10.6	Education spending, % of GDP	...
Doctors per 1,000 pop.	2.6	Enrolment, %: primary	101
Hospital beds per 1,000 pop.	2.7	secondary	114
At least basic drinking water,		tertiary	69
% of pop.	99.4		

Society

No. of households, m	14.1	Cost of living, Dec. 2019	
Av. no. per household	2.6	New York = 100	75
Marriages per 1,000 pop.	...	Cars per 1,000 pop.	612
Divorces per 1,000 pop.	...	Telephone lines per 100 pop.	37.3
Religion, % of pop.		Mobile telephone subscribers	
Christian	69.0	per 100 pop.	89.6
Non-religious	23.7	Internet access, %	91.0
Other	2.8	Broadband subs per 100 pop.	39.0
Muslim	2.1	Broadband speed, Mbps	19.5
Hindu	1.4		
Jewish	1.0		

a Including freshwater.

CHILE

Area, sq km	756,700	Capital	Santiago
Arable as % of total land	1.7	Currency	Chilean peso (Ps)

People

Population, m	18.7	Life expectancy: men	78.5 yrs
Pop. per sq km	24.7	women	82.8 yrs
Average annual rate of change		Adult literacy	96.4
in pop. 2015–20, %	1.2	Fertility rate (per woman)	1.7
Pop. aged 0–19, 2020, %	25.7	Urban population, %	87.6
Pop. aged 65 and over, 2020, %	12.2		per 1,000 pop.
No. of men per 100 women	97.3	Crude birth rate	12.5
Human Development Index	84.7	Crude death rate	6.6

The economy

GDP	$298bn	GDP per head	$15,923
GDP	Peso191trn	GDP per head in purchasing	
Av. ann. growth in real		power parity (USA=100)	40.2
GDP 2013–18	2.2%	Economic freedom index	76.8

Origins of GDP		**Components of GDP**	
	% of total		% of total
Agriculture	3.9	Private consumption	63.0
Industry, of which:	32.5	Public consumption	14.2
manufacturing	11.5	Investment	22.7
Services	63.6	Exports	28.8
		Imports	-28.7

Structure of employment

	% of total		% of labour force
Agriculture	9.0	Unemployed 2019	7.2
Industry	22.2	Av. ann. rate 2009–19	7.1
Services	68.8		

Energy

	m TOE		
Total output	10.5	Net energy imports as %	
Total consumption	38.1	of energy use	72
Consumption per head			
kg oil equivalent	2,111		

Inflation and finance

			% change 2018–19
Consumer price			
inflation 2019	2.3%	Narrow money (M1)	9.2
Av. ann. inflation 2014–19	3.0%	Broad money	12.4
Deposit rate, Dec. 2019	2.04%		

Exchange rates

	end 2019		December 2019
			2010 = 100
Ps per $	744.6	Effective rates	
Ps per SDR	1,029.7	– nominal	85.2
Ps per €	836.7	– real	85.0

Trade

Principal exports		Principal imports	
	$bn fob		*$bn cif*
Copper	28.0	Intermediate goods	37.0
Fresh fruit	5.7	Consumer goods	22.5
Salmon & trout	4.7	Capital goods	15.3
Cellulose & paper products	4.3		
Total incl. others	**75.3**	Total incl. others	**75.9**

Main export destinations		Main origins of imports	
	% of total		*% of total*
China	33.6	China	23.1
United States	13.9	United States	18.6
Japan	9.3	Brazil	8.9
South Korea	5.8	Argentina	4.5

Balance of payments, reserves and debt, $bn

Visible exports fob	75.5	Change in reserves	0.9
Visible imports fob	-70.8	Level of reserves	
Trade balance	4.7	end Dec.	39.9
Invisibles inflows	19.6	No. months of import cover	4.5
Invisibles outflows	-35.8	Official gold holdings, m oz	0.0
Net transfers	2.4	Foreign debt	185.5
Current account balance	-9.2	– as % of GDP	62.2
– as % of GDP	-3.1	– as % of total exports	199.1
Capital balance	9.5	Debt service ratio	44.6
Overall balance	1.4		

Health and education

Health spending, % of GDP	9.0	Education spending, % of GDP	5.4
Doctors per 1,000 pop.	2.6	Enrolment, %: primary	101
Hospital beds per 1,000 pop.	2.2	secondary	102
At least basic drinking water,		tertiary	88
% of pop.	99.8		

Society

No. of households, m	6.4	Cost of living, Dec. 2019	
Av. no. per household	2.9	New York = 100	64
Marriages per 1,000 pop.	3.3	Cars per 1,000 pop.	181
Divorces per 1,000 pop.	...	Telephone lines per 100 pop.	16.0
Religion, % of pop.		Mobile telephone subscribers	
Christian	89.4	per 100 pop.	134.4
Non-religious	8.6	Internet access, %	82.3
Other	1.9	Broadband subs per 100 pop.	17.4
Jewish	0.1	Broadband speed, Mbps	4.7
Hindu	<0.1		
Muslim	<0.1		

CHINA

Area, sq km	9,562,911	Capital	Beijing
Arable as % of total land	12.7	Currency	Yuan

People

Population, m	1,427.6	Life expectancy: men	75.4 yrs
Pop. per sq km	148.8	women	79.7 yrs
Average annual rate of change		Adult literacy	96.8
in pop. 2015–20, %	0.5	Fertility rate (per woman)	1.7
Pop. aged 0–19, 2020, %	23.4	Urban population, %	59.2
Pop. aged 65 and over, 2020, %	12.0		per 1,000 pop.
No. of men per 100 women	105.3	Crude birth rate	11.9
Human Development Index	75.8	Crude death rate	7.8

The economy

GDP	$13,608bn	GDP per head	$9,771
GDP	Yuan 88,443bn	GDP per head in purchasing	
Av. ann. growth in real		power parity (USA=100)	29.0
GDP 2013–18	6.9%	Economic freedom index	59.5

Origins of GDP		Components of GDP	
	% of total		% of total
Agriculture	7.4	Private consumption	38.7
Industry, of which:	39.9	Public consumption	14.7
manufacturing	27.8	Investment	44.1
Services	52.8	Exports	19.5
		Imports	-18.7

Structure of employment

	% of total		% of labour force
Agriculture	25.4	Unemployed 2019	4.3
Industry	28.2	Av. ann. rate 2009–19	4.3
Services	46.4		

Energy

	m TOE		
Total output	2,825.0	Net energy imports as %	
Total consumption	3,516.5	of energy use	20
Consumption per head			
kg oil equivalent	2,495		

Inflation and finance

			% change 2018–19
Consumer price			
inflation 2019	2.9%	Narrow money (M1)	11.8
Av. ann. inflation 2014–19	2.0%	Broad money	8.7
Deposit rate, Dec. 2019	1.50%		

Exchange rates

	end 2019		December 2019
Yuan per $	6.99	Effective rates	2010 = 100
Yuan per SDR	9.66	– nominal	114.8
Yuan per €	7.85	– real	120.6

Trade

Principal exports		Principal imports	
	$bn fob		*$bn cif*
Electrical goods	330.3	Electrical machinery	441.8
Telecoms equipment	326.8	Petroleum products	272.0
Office machinery	220.2	Metal ores & scrap	150.7
Clothing & apparel	158.9	Professional instruments	79.9
Total incl. others	**2,486.7**	Total incl. others	**2,135.7**

Main export destinations		Main origins of imports	
	% of total		*% of total*
United States	19.3	South Korea	9.5
Hong Kong	12.2	Taiwan	8.5
Japan	5.9	Japan	8.3
South Korea	4.4	United States	7.3
EU28	16.5	EU28	12.8

Balance of payments, reserves and debt, $bn

Visible exports fob	2,417.4	Change in reserves	-67.3
Visible imports fob	-2,022.3	Level of reserves	
Trade balance	395.2	end Dec.	3,168.1
Invisibles inflows	480.5	No. months of import cover	13.2
Invisibles outflows	-847.8	Official gold holdings, m oz	59.6
Net transfers	-2.4	Foreign debt	1,962.3
Current account balance	25.5	– as % of GDP	14.2
– as % of GDP	0.2	– as % of total exports	67.4
Capital balance	172.1	Debt service ratio	8.1
Overall balance	18.9		

Health and education

Health spending, % of GDP	5.2	Education spending, % of GDP	...
Doctors per 1,000 pop.	2.0	Enrolment, %: primary	100
Hospital beds per 1,000 pop.	4.2	secondary	...
At least basic drinking water,		tertiary	51
% of pop.	92.8		

Society

No. of households, m	473.8	Cost of living, Dec. 2019	
Av. no. per household	3.0	New York = 100	72
Marriages per 1,000 pop.	...	Cars per 1,000 pop.	108
Divorces per 1,000 pop.	...	Telephone lines per 100 pop.	13.5
Religion, % of pop.		Mobile telephone subscribers	
Non-religious	52.2	per 100 pop.	115.5
Other	22.7	Internet access, %	54.3
Buddhist	18.2	Broadband subs per 100 pop.	28.5
Christian	5.1	Broadband speed, Mbps	2.4
Muslim	1.8		
Jewish	<0.1		

Note: Data exclude Special Administrative Regions, ie, Hong Kong and Macau.

COLOMBIA

Area, sq km	1,141,749	Capital	Bogotá
Arable as % of total land	1.5	Currency	Colombian peso (peso)

People

Population, m	49.7	Life expectancy: men	75.2 yrs
Pop. per sq km	43.5	women	80.5 yrs
Average annual rate of change		Adult literacy	95.1
in pop. 2015–20, %	1.4	Fertility rate (per woman)	1.8
Pop. aged 0–19, 2020, %	30.5	Urban population, %	80.8
Pop. aged 65 and over, 2020, %	9.1		per 1,000 pop.
No. of men per 100 women	96.5	Crude birth rate	15.0
Human Development Index	76.1	Crude death rate	5.9

The economy

GDP	$331bn	GDP per head	$6,668
GDP	978trn peso	GDP per head in purchasing	
Av. ann. growth in real		power parity (USA=100)	23.9
GDP 2013–18	2.7%	Economic freedom index	69.2

Origins of GDP		**Components of GDP**	
	% of total		% of total
Agriculture	6.9	Private consumption	68.5
Industry, of which:	29.7	Public consumption	15.2
manufacturing	12.3	Investment	21.2
Services	63.4	Exports	15.9
		Imports	-20.8

Structure of employment

	% of total		% of labour force
Agriculture	16.6	Unemployed 2019	9.1
Industry	20.0	Av. ann. rate 2009–19	9.7
Services	63.4		

Energy

	m TOE		
Total output	129.3	Net energy imports as %	
Total consumption	42.1	of energy use	-207
Consumption per head			
kg oil equivalent	859		

Inflation and finance

		% change 2018–19	
Consumer price			
inflation 2019	3.5%	Narrow money (M1)	14.7
Av. ann. inflation 2014–19	4.7%	Broad money	8.9
Deposit rate, Dec. 2019	4.45%		

Exchange rates

	end 2019		December 2019
Peso per $	3,294.1	Effective rates	2010 = 100
Peso per SDR	4,555.1	– nominal	71.4
Peso per €	3,701.2	– real	71.2

Trade

Principal exports		Principal imports	
	$bn fob		*$bn cif*
Petroleum & products	16.8	Intermediate goods &	
Coal	7.4	raw materials	23.4
Coffee	2.3	Capital goods	15.8
Gold	1.5	Consumer goods	12.1
Total incl. others	**41.9**	Total	**51.2**

Main export destinations		Main origins of imports	
	% of total		*% of total*
United States	26.4	United States	25.6
Panama	9.2	China	20.6
China	8.5	Mexico	7.7
Ecuador	4.4	Brazil	5.5

Balance of payments, reserves and debt, $bn

Visible exports fob	44.4	Change in reserves	0.8
Visible imports fob	-49.6	Level of reserves	
Trade balance	-5.1	end Dec.	47.9
Invisibles inflows	15.8	No. months of import cover	7.1
Invisibles outflows	-31.3	Official gold holdings, m oz	0.4
Net transfers	7.6	Foreign debt	134.9
Current account balance	-13.0	– as % of GDP	40.5
– as % of GDP	-3.9	– as % of total exports	202.8
Capital balance	13.6	Debt service ratio	36.8
Overall balance	1.2		

Health and education

Health spending, % of GDP	7.2	Education spending, % of GDP	4.5
Doctors per 1,000 pop.	2.2	Enrolment, %: primary	115
Hospital beds per 1,000 pop.	1.5	secondary	98
At least basic drinking water,		tertiary	55
% of pop.	97.3		

Society

No. of households, m	14.4	Cost of living, Dec. 2019	
Av. no. per household	3.5	New York = 100	54
Marriages per 1,000 pop.	...	Cars per 1,000 pop.	68
Divorces per 1,000 pop.	...	Telephone lines per 100 pop.	14.0
Religion, % of pop.		Mobile telephone subscribers	
Christian	92.5	per 100 pop.	129.9
Non-religious	6.6	Internet access, %	64.1
Other	0.8	Broadband subs per 100 pop.	13.5
Hindu	<0.1	Broadband speed, Mbps	3.3
Jewish	<0.1		
Muslim	<0.1		

CZECH REPUBLIC

Area, sq km	78,870	Capital	Prague
Arable as % of total land	32.3	Currency	Koruna (Kc)

People

Population, m	10.7	Life expectancy: men	77.3 yrs
Pop. per sq km	135.7	women	82.4 yrs
Average annual rate of change		Adult literacy	...
in pop. 2015–20, %	0.2	Fertility rate (per woman)	1.6
Pop. aged 0–19, 2020, %	20.3	Urban population, %	73.8
Pop. aged 65 and over, 2020, %	20.1		per 1,000 pop.
No. of men per 100 women	97.0	Crude birth rate	10.5
Human Development Index	89.1	Crude death rate	10.9

The economy

GDP	$245bn	GDP per head	$23,079
GDP	Kc5,329bn	GDP per head in purchasing	
Av. ann. growth in real		power parity (USA=100)	63.3
GDP 2013–18	3.6%	Economic freedom index	74.8

Origins of GDP		**Components of GDP**	
	% of total		% of total
Agriculture	2.2	Private consumption	47.5
Industry, of which:	35.8	Public consumption	20.0
manufacturing	25.7	Investment	26.2
Services	62.0	Exports	78.4
		Imports	-72.0

Structure of employment

	% of total		% of labour force
Agriculture	2.7	Unemployed 2019	2.2
Industry	37.3	Av. ann. rate 2009-19	1.9
Services	60.0		

Energy

	m TOE		
Total output	26.3	Net energy imports as %	
Total consumption	44.9	of energy use	41
Consumption per head			
kg oil equivalent	4,225		

Inflation and finance

		% change 2018–19	
Consumer price			
inflation 2019	2.9%	Narrow money (M1)	4.5
Av. ann. inflation 2014–19	1.7%	Broad money	6.4
Deposit rate, Dec. 2019	0.43%		

Exchange rates

	end 2019		December 2019
			2010 = 100
Kc per $	22.62	Effective rates	
Kc per SDR	31.28	– nominal	101.3
Kc per €	25.42	– real	99.8

Trade

Principal exports	$bn fob	Principal imports	$bn cif
Machinery & transport equip.	117.9	Machinery & transport equip.	87.5
Semi-manufactures	30.2	Semi-manufactures	30.4
Miscellaneous manuf. goods	25.0	Miscellaneous manuf. goods	21.7
Chemicals	12.6	Chemicals	20.3
Total incl. others	**202.8**	Total incl. others	**185.1**

Main export destinations	% of total	Main origins of imports	% of total
Germany	32.4	Germany	29.0
Slovakia	7.5	Poland	9.1
Poland	6.1	China	8.4
France	5.1	Slovakia	6.0
EU28	84.4	EU28	76.5

Balance of payments, reserves and debt, $bn

Visible exports fob	161.1	Change in reserves	-5.5
Visible imports fob	-151.7	Level of reserves	
Trade balance	9.4	end Dec.	142.5
Invisibles inflows	42.4	No. months of import cover	8.5
Invisibles outflows	-48.7	Official gold holdings, m oz	0.3
Net transfers	-1.9	Foreign debt	194.1
Current account balance	1.2	– as % of GDP	79.2
– as % of GDP	0.5	– as % of total exports	93.8
Capital balance	-0.1	Debt service ratio	13.2
Overall balance	2.2		

Health and education

Health spending, % of GDP	7.2	Education spending, % of GDP	5.6
Doctors per 1,000 pop.	4.1	Enrolment, %: primary	101
Hospital beds per 1,000 pop.	6.5	secondary	103
At least basic drinking water,		tertiary	64
% of pop.	99.9		

Society

No. of households, m	4.4	Cost of living, Dec. 2019	
Av. no. per household	2.4	New York = 100	57
Marriages per 1,000 pop.	5.1	Cars per 1,000 pop.	499
Divorces per 1,000 pop.	2.3	Telephone lines per 100 pop.	14.2
Religion, % of pop.		Mobile telephone subscribers	
Non-religious	76.4	per 100 pop.	119.1
Christian	23.3	Internet access, %	80.7
Other	0.2	Broadband subs per 100 pop.	30.2
Hindu	<0.1	Broadband speed, Mbps	23.7
Jewish	<0.1		
Muslim	<0.1		

DENMARK

Area, sq km	42,920	Capital	Copenhagen
Arable as % of total land	56.5	Currency	Danish krone (DKr)

People

Population, m	5.8	Life expectancy: men	79.5 yrs
Pop. per sq km	135.1	women	83.3 yrs
Average annual rate of change		Adult literacy	...
in pop. 2015–20, %	0.4	Fertility rate (per woman)	1.8
Pop. aged 0–19, 2020, %	22.1	Urban population, %	87.9
Pop. aged 65 and over, 2020, %	20.2		per 1,000 pop.
No. of men per 100 women	98.8	Crude birth rate	10.7
Human Development Index	93.0	Crude death rate	10.0

The economy

GDP	$356bn	GDP per head	$61,350
GDP	DKr2,223bn	GDP per head in purchasing	
Av. ann. growth in real		power parity (USA=100)	88.7
GDP 2013–18	2.3%	Economic freedom index	78.3

Origins of GDP		**Components of GDP**	
	% of total		% of total
Agriculture	1.2	Private consumption	46.7
Industry, of which:	24.4	Public consumption	24.3
manufacturing	15.0	Investment	23.0
Services	74.4	Exports	55.6
		Imports	-49.6

Structure of employment

	% of total		% of labour force
Agriculture	2.2	Unemployed 2019	5.0
Industry	18.5	Av. ann. rate 2009-19	4.9
Services	79.3		

Energy

	m TOE		
Total output	17.2	Net energy imports as %	
Total consumption	18.6	of energy use	8
Consumption per head			
kg oil equivalent	3,251		

Inflation and finance

Consumer price		% change 2018–19	
inflation 2019	0.7%	Narrow money (M1)	2.0
Av. ann. inflation 2014–19	0.5%	Broad money	6.3
Money market rate, Dec. 2019	-0.59%		

Exchange rates

	end 2019		December 2019
DKr per $	6.68	Effective rates	2010 = 100
DKr per SDR	9.23	– nominal	101.7
DKr per €	7.51	– real	94.4

Trade

Principal exports		Principal imports	
	$bn fob		*$bn cif*
Machinery & transport equip.	28.8	Machinery & transport equip.	34.9
Chemicals & related products	23.4	Food, drink & tobacco	13.6
Food, drink & tobacco	19.6	Chemicals & related products	12.6
Mineral fuels & lubricants	6.0	Mineral fuels & lubricants	6.8
Total incl. others	**107.8**	Total incl. others	**101.4**

Main export destinations		Main origins of imports	
	% of total		*% of total*
Germany	16.1	Germany	22.8
Sweden	11.6	Sweden	11.8
United States	8.2	Netherlands	7.8
United Kingdom	7.1	China	7.1
EU28	61.2	EU28	70.1

Balance of payments, reserves and aid, $bn

Visible exports fob	119.6	Overall balance	-1.1
Visible imports fob	-105.3	Change in reserves	-4.3
Trade balance	14.2	Level of reserves	
Invisibles inflows	109.1	end Dec.	70.9
Invisibles outflows	-92.8	No. months of import cover	4.3
Net transfers	-5.7	Official gold holdings, m oz	2.1
Current account balance	24.8	Aid given	2.6
– as % of GDP	7.0	– as % of GNI	0.7
Capital balance	-14.0		

Health and education

Health spending, % of GDP	10.1	Education spending, % of GDP	...
Doctors per 1,000 pop.	4.0	Enrolment, %: primary	101
Hospital beds per 1,000 pop.	2.5	secondary	129
At least basic drinking water,		tertiary	81
% of pop.	100		

Society

No. of households, m	2.7	Cost of living, Dec. 2019	
Av. no. per household	2.1	New York = 100	92
Marriages per 1,000 pop.	5.6	Cars per 1,000 pop.	426
Divorces per 1,000 pop.	2.6	Telephone lines per 100 pop.	19.7
Religion, % of pop.		Mobile telephone subscribers	
Christian	83.5	per 100 pop.	125.1
Non-religious	11.8	Internet access, %	97.3
Muslim	4.1	Broadband subs per 100 pop.	44.1
Hindu	0.4	Broadband speed, Mbps	44.0
Other	0.2		
Jewish	<0.1		

EGYPT

Area, sq km	1,001,450	Capital	Cairo
Arable as % of total land	2.8	Currency	Egyptian pound (£E)

People

Population, m	98.4	Life expectancy: men	70.2 yrs
Pop. per sq km	98.2	women	75.0 yrs
Average annual rate of change		Adult literacy	71.2
in pop. 2015–20, %	2.0	Fertility rate (per woman)	3.3
Pop. aged 0–19, 2020, %	42.4	Urban population, %	42.7
Pop. aged 65 and over, 2020, %	5.3		per 1,000 pop.
No. of men per 100 women	102.1	Crude birth rate	26.5
Human Development Index	70.0	Crude death rate	5.7

The economy

GDP	$251bn	GDP per head	$2,549
GDP	£E4,437bn	GDP per head in purchasing	
Av. ann. growth in real		power parity (USA=100)	19.8
GDP 2013–18	4.2%	Economic freedom index	54.0

Origins of GDP		**Components of GDP**	
	% of total		% of total
Agriculture	11.5	Private consumption	85.4
Industry, of which:	34.7	Public consumption	8.4
manufacturing	16.7	Investment	16.7
Services	53.8	Exports	18.9
		Imports	-29.4

Structure of employment

	% of total		% of labour force
Agriculture	23.8	Unemployed 2019	11.6
Industry	27.7	Av. ann. rate 2009–19	10.8
Services	48.5		

Energy

	m TOE		
Total output	86.9	Net energy imports as %	
Total consumption	101.2	of energy use	14
Consumption per head			
kg oil equivalent	1,037		

Inflation and finance

Consumer price			% change 2018–19
inflation 2019	13.9%	Narrow money (M1)	7.7
Av. ann. inflation 2014–19	15.8%	Broad money	13.3
Deposit rate, Dec. 2019	9.37%		

Exchange rates

	end 2019		December 2019
£E per $	15.99	Effective rates	2010 = 100
£E per SDR	22.12	– nominal	...
£E per €	19.97	– real	...

Trade

Principal exports	$bn fob	Principal imports	$bn cif
Petroleum & products	8.8	Petroleum & products	12.5
Food	3.1	Machinery & equip.	10.1
Chemicals	2.7	Chemicals	5.5
Finished goods incl. textiles	2.3	Vehicles	2.9
Total incl. others	**28.0**	Total incl. others	**65.8**

Main export destinations	% of total	Main origins of imports	% of total
Italy	11.8	China	9.0
United States	8.4	Saudi Arabia	6.6
United Arab Emirates	8.0	United Arab Emirates	5.1
United Kingdom	6.3	United States	5.0

Balance of payments, reserves and debt, $bn

Visible exports fob	28.0	Change in reserves	5.4
Visible imports fob	-57.6	Level of reserves	
Trade balance	-29.6	end Dec.	41.8
Invisibles inflows	24.5	No. months of import cover	5.9
Invisibles outflows	-28.1	Official gold holdings, m oz	2.5
Net transfers	25.5	Foreign debt	98.7
Current account balance	-7.7	– as % of GDP	34.9
– as % of GDP	-3.1	– as % of total exports	126.5
Capital balance	14.5	Debt service ratio	10.1
Overall balance	3.2		

Health and education

Health spending, % of GDP	5.3	Education spending, % of GDP	...
Doctors per 1,000 pop.	0.5	Enrolment, %: primary	106
Hospital beds per 1,000 pop.	1.6	secondary	88
At least basic drinking water,		tertiary	35
% of pop.	99.1		

Society

No. of households, m	23.5	Cost of living, Dec. 2019	
Av. no. per household	4.2	New York = 100	52
Marriages per 1,000 pop.	9.6	Cars per 1,000 pop.	48
Divorces per 1,000 pop.	2.1	Telephone lines per 100 pop.	8.0
Religion, % of pop.		Mobile telephone subscribers	
Muslim	94.9	per 100 pop.	95.3
Christian	5.1	Internet access, %	46.9
Hindu	<0.1	Broadband subs per 100 pop.	6.7
Jewish	<0.1	Broadband speed, Mbps	1.3
Non-religious	<0.1		
Other	<0.1		

FINLAND

Area, sq km	338,450	Capital	Helsinki
Arable as % of total land	7.4	Currency	Euro (€)

People

Population, m	5.5	Life expectancy: men	79.8 yrs
Pop. per sq km	16.3	women	85.2 yrs
Average annual rate of change		Adult literacy	...
in pop. 2015–20, %	0.2	Fertility rate (per woman)	1.5
Pop. aged 0–19, 2020, %	21.2	Urban population, %	85.4
Pop. aged 65 and over, 2020, %	22.6		per 1,000 pop.
No. of men per 100 women	97.3	Crude birth rate	9.4
Human Development Index	92.5	Crude death rate	10.1

The economy

GDP	$277bn	GDP per head	$50,152
GDP	€232bn	GDP per head in purchasing	
Av. ann. growth in real		power parity (USA=100)	77.1
GDP 2013–18	1.5%	Economic freedom index	75.7

Origins of GDP		**Components of GDP**	
	% of total		% of total
Agriculture	2.8	Private consumption	52.8
Industry, of which:	27.8	Public consumption	22.7
manufacturing	17.0	Investment	24.9
Services	69.4	Exports	38.6
		Imports	-39.3

Structure of employment

	% of total		% of labour force
Agriculture	3.6	Unemployed 2019	7.4
Industry	22.2	Av. ann. rate 2009–19	6.6
Services	74.2		

Energy

	m TOE		
Total output	13.0	Net energy imports as %	
Total consumption	30.6	of energy use	58
Consumption per head			
kg oil equivalent	5,532		

Inflation and finance

		% change 2018–19	
Consumer price			
inflation 2019	1.1%	Narrow money (M1)	1.8
Av. ann. inflation 2014–19	0.7%	Broad money	5.2
Deposit rate, Dec. 2019	0.11%		

Exchange rates

	end 2019		December 2019
€ per $	0.89	Effective rates	2010 = 100
€ per SDR	1.23	– nominal	103.9
		– real	96.0

Trade

Principal exports		Principal imports	
	$bn fob		*$bn cif*
Machinery & transport equip.	23.8	Machinery & transport equip.	27.1
Raw materials	7.8	Mineral fuels & lubricants	12.4
Chemicals & related products	7.2	Chemicals & related products	9.0
Mineral fuels & lubricants	6.9	Food, drink & tobacco	5.5
Total incl. others	**74.3**	Total incl. others	**74.0**

Main export destinations		Main origins of imports	
	% of total		*% of total*
Germany	15.3	Germany	18.5
Sweden	10.3	Sweden	16.5
United States	7.1	Russia	14.7
Netherlands	6.8	Netherlands	9.1
EU28	59.0	EU28	70.2

Balance of payments, reserves and aid, $bn

Visible exports fob	74.3	Overall balance	-0.1
Visible imports fob	-74.0	Change in reserves	-0.2
Trade balance	0.3	Level of reserves	
Invisibles inflows	52.4	end Dec.	10.3
Invisibles outflows	-54.6	No. months of import cover	1.0
Net transfers	-2.8	Official gold holdings, m oz	1.6
Current account balance	-4.6	Aid given	1.0
– as % of GDP	-1.7	– as % of GNI	0.4
Capital balance	14.6		

Health and education

Health spending, % of GDP	9.2	Education spending, % of GDP	6.9
Doctors per 1,000 pop.	3.8	Enrolment, %: primary	100
Hospital beds per 1,000 pop.	4.4	secondary	154
At least basic drinking water,		tertiary	88
% of pop.	100		

Society

No. of households, m	2.7	Cost of living, Dec. 2019	
Av. no. per household	2.0	New York = 100	82
Marriages per 1,000 pop.	4.3	Cars per 1,000 pop.	480
Divorces per 1,000 pop.	2.4	Telephone lines per 100 pop.	5.9
Religion, % of pop.		Mobile telephone subscribers	
Christian	81.6	per 100 pop.	129.5
Non-religious	17.6	Internet access, %	88.9
Muslim	0.8	Broadband subs per 100 pop.	31.5
Hindu	<0.1	Broadband speed, Mbps	24.0
Jewish	<0.1		
Other	<0.1		

FRANCE

Area, sq km	549,087	Capital	Paris
Arable as % of total land	33.5	Currency	Euro (€)

People

Population, m	65.0	Life expectancy: men	80.3 yrs
Pop. per sq km	117.9	women	85.8 yrs
Average annual rate of change		Adult literacy	...
in pop. 2015–20, %	0.3	Fertility rate (per woman)	1.9
Pop. aged 0–19, 2020, %	23.6	Urban population, %	80.4
Pop. aged 65 and over, 2020, %	20.8		per 1,000 pop.
No. of men per 100 women	93.8	Crude birth rate	11.2
Human Development Index	89.1	Crude death rate	9.5

The economy

GDP	$2,778bn	GDP per head	$41,464
GDP	€2,353bn	GDP per head in purchasing	
Av. ann. growth in real		power parity (USA=100)	72.2
GDP 2013–18	1.4%	Economic freedom index	66.0

Origins of GDP		**Components of GDP**	
	% of total		% of total
Agriculture	1.8	Private consumption	53.9
Industry, of which:	19.0	Public consumption	23.4
manufacturing	10.9	Investment	23.5
Services	79.2	Exports	31.3
		Imports	-32.1

Structure of employment

	% of total		% of labour force
Agriculture	2.4	Unemployed 2019	9.1
Industry	20.1	Av. ann. rate 2009-19	8.4
Services	77.5		

Energy

	m TOE		
Total output	122.9	Net energy imports as %	
Total consumption	260.3	of energy use	53
Consumption per head			
kg oil equivalent	4,006		

Inflation and finance

		% change 2018–19	
Consumer price			
inflation 2019	1.3%	Narrow money (M1)	1.8
Av. ann. inflation 2014–19	1.0%	Broad money	5.2
Deposit rate, Dec. 2019	0.78%		

Exchange rates

	end 2019		December 2019
€ per $	0.89	Effective rates	2010 = 100
€ per SDR	1.23	– nominal	100.8
		– real	93.3

Trade

Principal exports	$bn fob	Principal imports	$bn cif
Machinery & transport equip.	226.8	Machinery & transport equip.	239.8
Chemicals & related products	108.3	Chemicals & related products	88.9
Food, drink and tobacco	68.0	Mineral fuels & lubricants	73.8
Mineral fuels & lubricants	20.7	Food, drink and tobacco	59.6
Total incl. others	**581.3**	**Total incl. others**	**655.3**

Main export destinations	% of total	Main origins of imports	% of total
Germany	14.5	Germany	18.8
United States	7.9	Belgium	10.5
Spain	7.8	Netherlands	8.4
Italy	7.4	Italy	8.3
EU28	59.0	EU28	69.0

Balance of payments, reserves and aid, $bn

Visible exports fob	611.1	Overall balance	12.4
Visible imports fob	-669.7	Change in reserves	10.4
Trade balance	-58.7	Level of reserves	
Invisibles inflows	509.7	end Dec.	166.3
Invisibles outflows	-413.6	No. months of import cover	1.8
Net transfers	-56.5	Official gold holdings, m oz	78.3
Current account balance	-19.0	Aid given	12.8
– as % of GDP	-0.7	– as % of GNI	0.4
Capital balance	49.2		

Health and education

Health spending, % of GDP	11.3	Education spending, % of GDP	...
Doctors per 1,000 pop.	3.3	Enrolment, %: primary	103
Hospital beds per 1,000 pop.	6.5	secondary	104
At least basic drinking water,		tertiary	66
% of pop.	100		

Society

No. of households, m	29.8	Cost of living, Dec. 2019	
Av. no. per household	2.2	New York = 100	99
Marriages per 1,000 pop.	3.5	Cars per 1,000 pop.	497
Divorces per 1,000 pop.	1.9	Telephone lines per 100 pop.	59.4
Religion, % of pop.		Mobile telephone subscribers	
Christian	63.0	per 100 pop.	108.4
Non-religious	28.0	Internet access, %	82.0
Muslim	7.5	Broadband subs per 100 pop.	44.8
Other	1.0	Broadband speed, Mbps	24.2
Jewish	0.5		
Hindu	<0.1		

GERMANY

Area, sq km	357,580	Capital	Berlin
Arable as % of total land	33.7	Currency	Euro (€)

People

Population, m	83.1	Life expectancy: men	79.6 yrs
Pop. per sq km	232.6	women	84.1 yrs
Average annual rate of change		Adult literacy	..
in pop. 2015–20, %	0.5	Fertility rate (per woman)	1.6
Pop. aged 0–19, 2020, %	18.9	Urban population, %	77.3
Pop. aged 65 and over, 2020, %	21.7		per 1,000 pop.
No. of men per 100 women	97.8	Crude birth rate	9.4
Human Development Index	93.9	Crude death rate	11.7

The economy

GDP	$3,948bn	GDP per head	$47,603
GDP	€3,344bn	GDP per head in purchasing	
Av. ann. growth in real		power parity (USA=100)	84.5
GDP 2013–18	2.0%	Economic freedom index	73.5

Origins of GDP		**Components of GDP**	
	% of total		% of total
Agriculture	0.9	Private consumption	52.1
Industry, of which:	30.5	Public consumption	19.9
manufacturing	22.7	Investment	21.8
Services	68.7	Exports	47.4
		Imports	-41.3

Structure of employment

	% of total		% of labour force
Agriculture	1.2	Unemployed 2019	3.4
Industry	27.0	Av. ann. rate 2009–19	3.0
Services	71.7		

Energy

	m TOE		
Total output	120.2	Net energy imports as %	
Total consumption	353.4	of energy use	66
Consumption per head			
kg oil equivalent	4,303		

Inflation and finance

		% change 2018–19	
Consumer price			
inflation 2019	1.3%	Narrow money (M1)	1.8
Av. ann. inflation 2014–19	1.2%	Broad money	5.2
Deposit rate, Dec. 2019	0.14%		

Exchange rates

	end 2019		December 2019
€ per $	0.89	Effective rates	2010 = 100
€ per SDR	1.23	– nominal	101.1
		– real	94.6

Trade

Principal exports	$bn fob	Principal imports	$bn cif
Machinery & transport equip.	771.1	Machinery & transport equip.	481.6
Chemicals & related products	255.4	Chemicals & related products	186.3
Food, drink and tobacco	79.1	Mineral fuels & lubricants	115.4
Mineral fuels & lubricants	37.5	Food, drink and tobacco	92.5
Total incl. others	**1,563.0**	Total incl. others	**1,289.9**

Main export destinations	% of total	Main origins of imports	% of total
United States	8.7	Netherlands	14.0
France	8.0	China	6.9
China	7.1	France	6.3
Netherlands	6.9	Belgium	6.1
EU28	59.0	EU28	66.4

Balance of payments, reserves and aid, $bn

Visible exports fob	1,527.7	Overall balance	0.5
Visible imports fob	-1,259.5	Change in reserves	-1.6
Trade balance	268.2	Level of reserves	
Invisibles inflows	616.7	end Dec.	197.7
Invisibles outflows	-534.6	No. months of import cover	1.3
Net transfers	-57.2	Official gold holdings, m oz	108.3
Current account balance	293.1	Aid given	25.7
– as % of GDP	7.4	– as % of GNI	0.7
Capital balance	-279.9		

Health and education

Health spending, % of GDP	11.2	Education spending, % of GDP	4.8
Doctors per 1,000 pop.	4.2	Enrolment, %: primary	104
Hospital beds per 1,000 pop.	8.3	secondary	98
At least basic drinking water,		tertiary	70
% of pop.	100		

Society

No. of households, m	40.8	Cost of living, Dec. 2019	
Av. no. per household	2.0	New York = 100	71
Marriages per 1,000 pop.	5.0	Cars per 1,000 pop.	542
Divorces per 1,000 pop.	1.8	Telephone lines per 100 pop.	51.1
Religion, % of pop.		Mobile telephone subscribers	
Christian	68.7	per 100 pop.	129.3
Non-religious	24.7	Internet access, %	89.7
Muslim	5.8	Broadband subs per 100 pop.	41.1
Other	0.5	Broadband speed, Mbps	24.0
Jewish	0.3		
Hindu	<0.1		

GREECE

Area, sq km	131,960	Capital	Athens
Arable as % of total land	16.6	Currency	Euro (€)

People

Population, m	10.5	Life expectancy: men	80.5 yrs
Pop. per sq km	79.6	women	85.1 yrs
Average annual rate of change		Adult literacy	97.9
in pop. 2015–20, %	-0.5	Fertility rate (per woman)	1.3
Pop. aged 0–19, 2020, %	18.7	Urban population, %	79.1
Pop. aged 65 and over, 2020, %	22.3		per 1,000 pop.
No. of men per 100 women	96.4	Crude birth rate	7.8
Human Development Index	87.2	Crude death rate	11.4

The economy

GDP	$218bn	GDP per head	$20,324
GDP	€185bn	GDP per head in purchasing	
Av. ann. growth in real		power parity (USA=100)	47.1
GDP 2013–18	0.7%	Economic freedom index	59.9

Origins of GDP		**Components of GDP**	
	% of total		% of total
Agriculture	4.3	Private consumption	68.0
Industry, of which:	17.5	Public consumption	19.1
manufacturing	11.0	Investment	13.1
Services	78.2	Exports	36.1
		Imports	-36.4

Structure of employment

	% of total		% of labour force
Agriculture	12.0	Unemployed 2019	19.3
Industry	15.2	Av. ann. rate 2009-19	17.2
Services	72.8		

Energy

	m TOE		
Total output	8.2	Net energy imports as %	
Total consumption	29.4	of energy use	72
Consumption per head			
kg oil equivalent	2,635		

Inflation and finance

			% change 2018–19
Consumer price			
inflation 2019	0.5%	Narrow money (M1)	1.8
Av. ann. inflation 2014–19	0.3%	Broad money	5.2
Deposit rate, Dec. 2019	0.37%		

Exchange rates

	end 2019		December 2019
€ per $	0.89	Effective rates	2010 = 100
€ per SDR	1.23	– nominal	104.7
		– real	88.3

Trade

Principal exports

	$bn fob
Mineral fuels & lubricants	13.6
Food, drink and tobacco	6.3
Chemicals & related products	4.1
Machinery & transport equip.	3.5
Total incl. others	**39.5**

Principal imports

	$bn cif
Machinery & transport equip.	18.9
Mineral fuels & lubricants	11.4
Chemicals & related products	9.5
Food, drink and tobacco	7.6
Total incl. others	**63.9**

Main export destinations

	% of total
Italy	10.4
Germany	6.4
Turkey	6.1
Cyprus	5.7
EU28	52.8

Main origins of imports

	% of total
Germany	10.7
Iraq	8.4
Italy	8.4
Russia	7.6
EU28	52.5

Balance of payments, reserves and debt, $bn

Visible exports fob	38.2	Overall balance	2.0
Visible imports fob	-64.7	Change in reserves	-0.2
Trade balance	-26.5	Level of reserves	
Invisibles inflows	50.9	end Dec.	7.6
Invisibles outflows	-30.3	No. months of import cover	1.0
Net transfers	-0.3	Official gold holdings, m oz	3.6
Current account balance	-6.2	Aid given	0.3
– as % of GDP	-2.9	– as % of GNI	0.2
Capital balance	7.3		

Health and education

Health spending, % of GDP	8.0	Education spending, % of GDP	...
Doctors per 1,000 pop.	5.5	Enrolment, %: primary	100
Hospital beds per 1,000 pop.	4.3	secondary	104
At least basic drinking water,		tertiary	137
% of pop.	100		

Society

No. of households, m	4.2	Cost of living, Dec. 2019	
Av. no. per household	2.5	New York = 100	58
Marriages per 1,000 pop.	4.4	Cars per 1,000 pop.	594
Divorces per 1,000 pop.	1.8	Telephone lines per 100 pop.	48.3
Religion, % of pop.		Mobile telephone subscribers	
Christian	88.1	per 100 pop.	115.7
Non-religious	6.1	Internet access, %	73.0
Muslim	5.3	Broadband subs per 100 pop.	37.7
Other	0.3	Broadband speed, Mbps	11.3
Hindu	0.1		
Jewish	<0.1		

HONG KONG

Area, sq km	1,110	Capital	Victoria
Arable as % of total land	2.9	Currency	Hong Kong dollar (HK$)

People

Population, m	7.4	Life expectancy: men	82.4 yrs
Pop. per sq km	6,883.7	women	88.2 yrs
Average annual rate of change		Adult literacy	...
in pop. 2015–20, %	0.9	Fertility rate (per woman)	1.3
Pop. aged 0–19, 2020, %	16.2	Urban population, %	100.0
Pop. aged 65 and over, 2020, %	18.2		per 1,000 pop.
No. of men per 100 women	84.8	Crude birth rate	11.1
Human Development Index	93.9	Crude death rate	7.2

The economy

GDP	$363bn	GDP per head	$48,676
GDP	HK$2,843bn	GDP per head in purchasing	
Av. ann. growth in real		power parity (USA=100)	102.9
GDP 2013–18	2.8%	Economic freedom index	89.1

Origins of GDP		**Components of GDP**	
	% of total		% of total
Agriculture	0.1	Private consumption	68.3
Industry, of which:	6.8	Public consumption	9.9
manufacturing	1.0	Investment	21.7
Services	93.1	Exports	188.3
		Imports	-188.2

Structure of employment

	% of total		% of labour force
Agriculture	0.2	Unemployed 2019	2.9
Industry	11.6	Av. ann. rate 2009-19	3.6
Services	88.2		

Energy

	m TOE		
Total output	0.0	Net energy imports as %	
Total consumption	33.3	of energy use	100
Consumption per head			
kg oil equivalent	4,526		

Inflation and finance

			% change 2018–19
Consumer price			
inflation 2019	2.9%	Narrow money (M1)	2.5
Av. ann. inflation 2014–19	2.4%	Broad money	2.8
Central bank policy rate, Dec. 2019	2.49%		

Exchange rates

	end 2019		December 2019
			2010 = 100
HK$ per $	7.79	Effective rates	106.1
HK$ per SDR	10.77	– nominal	106.1
HK$ per €	8.75	– real	...

Trade

Principal exports[a]	$bn fob	Principal imports[a]	$bn cif
Capital goods	211.7	Raw materials &	
Semi-finished goods	208.0	semi-manufactures	237.4
Consumer goods	95.7	Capital goods & raw materials	207.8
Foodstuffs	8.8	Consumer goods	116.2
		Foodstuffs	25.3
Total incl. others	**531.2**	Total incl. others	**602.9**

Main export destinations	% of total	Main origins of imports	% of total
China	55.0	China	46.3
United States	8.6	Taiwan	7.2
India	3.2	Singapore	6.6
Japan	3.1	South Korea	5.9

Balance of payments, reserves and debt, $bn

Visible exports fob	568.1	Change in reserves	-6.8
Visible imports fob	-600.4	Level of reserves	
Trade balance	-32.3	end Dec.	424.6
Invisibles inflows	321.1	No. months of import cover	5.8
Invisibles outflows	-272.4	Official gold holdings, m oz	0.1
Net transfers	-2.9	Foreign debt	691.6
Current account balance	13.5	– as % of GDP	190.7
– as % of GDP	3.7	– as % of total exports	77.5
Capital balance	-21.3	Debt service ratio	8.3
Overall balance	1.0		

Health and education

Health spending, % of GDP	...	Education spending, % of GDP	3.3
Doctors per 1,000 pop.	...	Enrolment, %: primary	109
Hospital beds per 1,000 pop.	...	secondary	107
At least basic drinking water,		tertiary	77
% of pop.	100		

Society

No. of households, m	2.5	Cost of living, Dec. 2019	
Av. no. per household	3.0	New York = 100	102
Marriages per 1,000 pop.	6.6	Cars per 1,000 pop.	74
Divorces per 1,000 pop.	...	Telephone lines per 100 pop.	56.9
Religion, % of pop.		Mobile telephone subscribers	
Non-religious	56.1	per 100 pop.	270.0
Christian	14.3	Internet access, %	90.5
Other	14.2	Broadband subs per 100 pop.	36.8
Buddhist	13.2	Broadband speed, Mbps	26.5
Muslim	1.8		
Hindu	0.4		

a Including re-exports.
Note: Hong Kong became a Special Administrative Region of China on July 1 1997.

HUNGARY

Area, sq km	93,030	Capital	Budapest
Arable as % of total land	47.8	Currency	Forint (Ft)

People

Population, m	9.7	Life expectancy: men	73.8 yrs
Pop. per sq km	104.3	women	80.7 yrs
Average annual rate of change		Adult literacy	99.1
in pop. 2015–20, %	-0.2	Fertility rate (per woman)	1.5
Pop. aged 0–19, 2020, %	19.4	Urban population, %	71.4
Pop. aged 65 and over, 2020, %	20.2		per 1,000 pop.
No. of men per 100 women	90.8	Crude birth rate	9.5
Human Development Index	84.5	Crude death rate	13.0

The economy

GDP	$158bn	GDP per head	$16,162
GDP	Ft42,073bn	GDP per head in purchasing	
Av. ann. growth in real		power parity (USA=100)	49.5
GDP 2013–18	3.9%	Economic freedom index	66.4

Origins of GDP		**Components of GDP**	
	% of total		% of total
Agriculture	4.2	Private consumption	48.7
Industry, of which:	30.1	Public consumption	19.7
manufacturing	22.1	Investment	27.2
Services	65.7	Exports	84.9
		Imports	-80.6

Structure of employment

	% of total		% of labour force
Agriculture	4.7	Unemployed 2019	3.7
Industry	32.7	Av. ann. rate 2009–19	3.4
Services	62.6		

Energy

	m TOE		
Total output	8.7	Net energy imports as %	
Total consumption	26.3	of energy use	67
Consumption per head			
kg oil equivalent	2,704		

Inflation and finance

			% change 2018–19
Consumer price			
inflation 2019	3.4%	Narrow money (M1)	16.7
Av. ann. inflation 2014–19	1.8%	Broad money	8.2
Deposit rate, Dec. 2019	0.13%		

Exchange rates

	end 2019		December 2019
			2010 = 100
Ft per $	294.74	Effective rates	
Ft per SDR	407.57	– nominal	85.5
Ft per €	331.17	– real	87.8

Trade

Principal exports		Principal imports	
	$bn fob		*$bn cif*
Machinery & equipment	69.0	Machinery & equipment	55.9
Manufactured goods	40.1	Manufactured goods	43.1
Food, drink & tobacco	8.4	Fuels & energy	9.6
Raw materials	2.8	Food, drink & tobacco	6.1
Total incl. others	**123.6**	Total incl. others	**117.2**

Main export destinations		Main origins of imports	
	% of total		*% of total*
Germany	27.9	Germany	26.6
Romania	5.4	Austria	6.5
Slovakia	5.3	China	6.5
Italy	5.2	Poland	5.8
EU28	81.8	EU28	74.8

Balance of payments, reserves and debt, $bn

Visible exports fob	104.7	Change in reserves	3.4
Visible imports fob	-106.7	Level of reserves	
Trade balance	-1.9	end Dec.	31.4
Invisibles inflows	44.9	No. months of import cover	2.5
Invisibles outflows	-41.8	Official gold holdings, m oz	1.0
Net transfers	-1.0	Foreign debt	152.7
Current account balance	0.1	– as % of GDP	96.9
– as % of GDP	0.1	– as % of total exports	101.5
Capital balance	6.4	Debt service ratio	36.0
Overall balance	4.1		

Health and education

Health spending, % of GDP	6.9	Education spending, % of GDP	4.7
Doctors per 1,000 pop.	3.4	Enrolment, %: primary	101
Hospital beds per 1,000 pop.	7.0	secondary	103
At least basic drinking water,		tertiary	49
% of pop.	100		

Society

No. of households, m	4.1	Cost of living, Dec. 2019	
Av. no. per household	2.4	New York = 100	51
Marriages per 1,000 pop.	5.2	Cars per 1,000 pop.	332
Divorces per 1,000 pop.	1.7	Telephone lines per 100 pop.	31.1
Religion, % of pop.		Mobile telephone subscribers	
Christian	81.0	per 100 pop.	103.5
Non-religious	18.6	Internet access, %	76.1
Other	0.2	Broadband subs per 100 pop.	31.7
Jewish	0.1	Broadband speed, Mbps	34.0
Hindu	<0.1		
Muslim	<0.1		

INDIA

Area, sq km	3,287,263	Capital	New Delhi
Arable as % of total land	52.6	Currency	Indian rupee (Rs)

People

Population, m	1,352.6	Life expectancy: men	69.2 yrs
Pop. per sq km	411.5	women	71.8 yrs
Average annual rate of change		Adult literacy	74.4
in pop. 2015–20, %	1.0	Fertility rate (per woman)	2.2
Pop. aged 0–19, 2020, %	35.3	Urban population, %	34.0
Pop. aged 65 and over, 2020, %	6.6		per 1,000 pop.
No. of men per 100 women	108.2	Crude birth rate	18.0
Human Development Index	64.7	Crude death rate	7.4

The economy

GDP	$2,719bn	GDP per head	$2,010
GDP	Rs190,102bn	GDP per head in purchasing	
Av. ann. growth in real		power parity (USA=100)	12.4
GDP 2013–18	7.5%	Economic freedom index	56.5

Origins of GDP		Components of GDP	
	% of total		% of total
Agriculture	17.2	Private consumption	59.4
Industry, of which:	29.2	Public consumption	11.2
manufacturing	16.4	Investment	31.3
Services	53.5	Exports	19.7
		Imports	-23.6

Structure of employment

	% of total		% of labour force
Agriculture	42.4	Unemployed 2019	5.3
Industry	25.6	Av. ann. rate 2009–19	5.4
Services	32.0		

Energy

	m TOE		
Total output	445.8	Net energy imports as %	
Total consumption	768.6	of energy use	42
Consumption per head			
kg oil equivalent	574		

Inflation and finance

		% change 2018–19	
Consumer price			
inflation 2019	4.5%	Narrow money (M1)	13.2
Av. ann. inflation 2014–19	4.2%	Broad money	10.5
Money market rate, Dec. 2019	5.06%		

Exchange rates

	end 2019		December 2019
Rs per $	71.27	Effective rates	2010 = 100
Rs per SDR	98.56	– nominal	...
Rs per €	80.08	– real	...

Trade

Principal exports

	$bn fob
Engineering products	83.6
Petroleum & products	46.5
Gems & jewellery	40.3
Agricultural products	33.0
Total incl. others	**330.1**

Principal imports

	$bn cif
Petroleum & products	140.9
Electronic goods	55.5
Machinery	39.8
Gold & silver	36.7
Total incl. others	**514.0**

Main export destinations

	% of total
United States	15.6
United Arab Emirates	8.8
China	5.0
Hong Kong	4.0

Main origins of imports

	% of total
China	13.8
United States	6.2
Saudi Arabia	5.3
United Arab Emirates	5.0

Balance of payments, reserves and debt, $bn

Visible exports fob	332.1	Change in reserves	-13.4
Visible imports fob	-518.8	Level of reserves	
Trade balance	-186.7	end Dec.	399.1
Invisibles inflows	226.3	No. months of import cover	6.9
Invisibles outflows	-175.3	Official gold holdings, m oz	19.3
Net transfers	70.1	Foreign debt	521.4
Current account balance	-65.6	– as % of GDP	19.2
– as % of GDP	-2.4	– as % of total exports	81.8
Capital balance	60.1	Debt service ratio	10.0
Overall balance	-3.8		

Health and education

Health spending, % of GDP	3.5	Education spending, % of GDP	...
Doctors per 1,000 pop.	0.9	Enrolment, %: primary	113
Hospital beds per 1,000 pop.	0.7	secondary	75
At least basic drinking water,		tertiary	28
% of pop.	92.7		

Society

No. of households, m	282.2	Cost of living, Dec. 2019	
Av. no. per household	4.8	New York = 100	42
Marriages per 1,000 pop.	...	Cars per 1,000 pop.	18
Divorces per 1,000 pop.	...	Telephone lines per 100 pop.	1.6
Religion, % of pop.		Mobile telephone subscribers	
Hindu	79.5	per 100 pop.	86.9
Muslim	14.4	Internet access, %	34.5
Other	3.6	Broadband subs per 100 pop.	1.3
Christian	2.5	Broadband speed, Mbps	5.2
Jewish	<0.1		
Non-religious	<0.1		

INDONESIA

Area, sq km	1,913,580	Capital	Jakarta
Arable as % of total land	13.0	Currency	Rupiah (Rp)

People

Population, m	267.7	Life expectancy: men	70.1 yrs
Pop. per sq km	140.1	women	74.6 yrs
Average annual rate of change		Adult literacy	95.7
in pop. 2015–20, %	1.1	Fertility rate (per woman)	2.3
Pop. aged 0–19, 2020, %	34.5	Urban population, %	55.3
Pop. aged 65 and over, 2020, %	6.3		per 1,000 pop.
No. of men per 100 women	101.4	Crude birth rate	18.2
Human Development Index	70.7	Crude death rate	6.8

The economy

GDP	$1,042bn	GDP per head	$3,894
GDP	Rs14,825,944bn	GDP per head in purchasing	
Av. ann. growth in real		power parity (USA=100)	20.8
GDP 2013–18	5.0%	Economic freedom index	67.2

Origins of GDP		**Components of GDP**	
	% of total		% of total
Agriculture	13.4	Private consumption	57.0
Industry, of which:	41.3	Public consumption	9.0
manufacturing	20.7	Investment	34.6
Services	45.2	Exports	21.0
		Imports	-22.1

Structure of employment

	% of total		% of labour force
Agriculture	28.6	Unemployed 2019	4.5
Industry	22.5	Av. ann. rate 2009-19	4.7
Services	48.9		

Energy

	m TOE		
Total output	359.1	Net energy imports as %	
Total consumption	180.5	of energy use	-99
Consumption per head			
kg oil equivalent	684		

Inflation and finance

		% change 2018–19	
Consumer price			
inflation 2019	2.8%	Narrow money (M1)	2.9
Av. ann. inflation 2014–19	4.0%	Broad money	6.5
Deposit rate, Dec. 2019	6.38%		

Exchange rates

	end 2019		December 2019
Rp per $	13,901.0	Effective rates	2010 = 100
Rp per SDR	19,222.7	– nominal	...
Rp per €	15,619.1	– real	...

Trade

Principal exports		Principal imports	
	$bn fob		*$bn cif*
Manufactured goods	126.8	Raw materials & auxiliary	
Mining & other sector products	44.1	materials	130.7
Agricultural goods	5.8	Capital goods	30.7
Unclassified exports	1.9	Consumer goods	26.0
Total incl. others	**180.0**	Total incl. others	**188.7**

Main export destinations		Main origins of imports	
	% of total		*% of total*
China	15.1	China	24.1
Japan	10.8	Singapore	11.4
United States	10.3	Japan	9.5
India	7.6	Thailand	5.8

Balance of payments, reserves and debt, $bn

Visible exports fob	180.7	Change in reserves	-9.5
Visible imports fob	-181.0	Level of reserves	
Trade balance	-0.2	end Dec.	120.7
Invisibles inflows	40.5	No. months of import cover	5.6
Invisibles outflows	-77.8	Official gold holdings, m oz	2.5
Net transfers	6.9	Foreign debt	369.8
Current account balance	-30.6	– as % of GDP	35.5
– as % of GDP	-2.9	– as % of total exports	159.1
Capital balance	25.2	Debt service ratio	11.7
Overall balance	-7.1		

Health and education

Health spending, % of GDP	3.0	Education spending, % of GDP	3.6
Doctors per 1,000 pop.	0.4	Enrolment, %: primary	106
Hospital beds per 1,000 pop.	1.2	secondary	89
At least basic drinking water,		tertiary	36
% of pop.	89.3		

Society

No. of households, m	66.9	Cost of living, Dec. 2019	
Av. no. per household	4.0	New York = 100	59
Marriages per 1,000 pop.	...	Cars per 1,000 pop.	55
Divorces per 1,000 pop.	...	Telephone lines per 100 pop.	3.1
Religion, % of pop.		Mobile telephone subscribers	
Muslim	87.2	per 100 pop.	119.3
Christian	9.9	Internet access, %	39.9
Hindu	1.7	Broadband subs per 100 pop.	3.3
Other	1.1	Broadband speed, Mbps	5.8
Jewish	<0.1		
Non-religious	<0.1		

IRAN

Area, sq km	1,745,150	Capital	Tehran
Arable as % of total land	9.0	Currency	Rial (IR)

People

Population, m	81.8	Life expectancy: men	76.2 yrs
Pop. per sq km	50.2	women	78.5 yrs
Average annual rate of change		Adult literacy	85.5
in pop. 2015–20, %	1.4	Fertility rate (per woman)	2.2
Pop. aged 0–19, 2020, %	31.3	Urban population, %	74.9
Pop. aged 65 and over, 2020, %	6.6		per 1,000 pop.
No. of men per 100 women	102.0	Crude birth rate	19.1
Human Development Index	79.7	Crude death rate	4.8

The economy

GDP	$446bn	GDP per head	$5,417
GDP	IR18,689,543bn	GDP per head in purchasing	
Av. ann. growth in real		power parity (USA=100)	30.9
GDP 2013–18	5.0%	Economic freedom index	49.2

Origins of GDP[a]		Components of GDP[a]	
	% of total		% of total
Agriculture	9.8	Private consumption	47.6
Industry, of which:	36.1	Public consumption	13.4
manufacturing	12.4	Investment	34.7
Services	54.0	Exports	24.9
		Imports	-23.8

Structure of employment

	% of total		% of labour force
Agriculture	17.9	Unemployed 2019	12.0
Industry	30.6	Av. ann. rate 2009–19	11.4
Services	51.5		

Energy

	m TOE		
Total output	459.2	Net energy imports as %	
Total consumption	292.5	of energy use	-57
Consumption per head			
kg oil equivalent	3,604		

Inflation and finance

		% change 2018–19	
Consumer price			
inflation 2019	41.1%	Narrow money (M1)	12.6
Av. ann. inflation 2014–19	19.9%	Broad money	28.2
Deposit rate, Feb. 2017	12.70%		

Exchange rates

	end 2019		December 2019
IR per $	42,000.00	Effective rates	2010 = 100
IR per SDR	58,078.7	– nominal	29.5
IR per €	47,191.0	– real	137.2

Trade

Principal exports[a]

	$bn fob
Oil & gas	65.8
Petrochemicals	9.0
Fresh & dry fruits	2.3
Carpets	0.4
Total incl. others	**98.1**

Principal imports[a]

	$bn fob
Machinery & transport equip.	21.9
Intermediate goods	7.8
Foodstuffs	7.6
Chemicals	7.2
Total incl. others	**75.5**

Main export destinations

	% of total
China	29.0
India	19.5
Turkey	9.1
United Arab Emirates	6.6

Main origins of imports

	% of total
China	22.6
United Arab Emirates	14.8
Germany	6.7
India	5.6

Balance of payments[b], reserves and debt, $bn

Visible exports fob	93.4	Change in reserves	2.4
Visible imports fob	-60.8	Level of reserves	
Trade balance	32.6	end Dec.	108.1
Invisibles inflows	...	No. months of import cover	19.1
Invisibles outflows	...	Official gold holdings, m oz	...
Net transfers	0.7	Foreign debt	6.3
Current account balance	26.7	– as % of GDP	1.5
– as % of GDP	6.0	– as % of total exports	5.8
Capital balance	...	Debt service ratio	0.9
Overall balance	...		

Health and education

Health spending, % of GDP	8.7	Education spending, % of GDP	4.0
Doctors per 1,000 pop.	1.6	Enrolment, %: primary	111
Hospital beds per 1,000 pop.	1.5	secondary	86
At least basic drinking water,		tertiary	68
% of pop.	95.2		

Society

No. of households, m	24.7	Cost of living, Dec. 2019	
Av. no. per household	3.3	New York = 100	52
Marriages per 1,000 pop.	7.5	Cars per 1,000 pop.	169
Divorces per 1,000 pop.	2.2	Telephone lines per 100 pop.	37.3
Religion, % of pop.		Mobile telephone subscribers	
Muslim	99.5	per 100 pop.	108.5
Christian	0.2	Internet access, %	70.0
Other	0.2	Broadband subs per 100 pop.	12.0
Non-religious	0.1	Broadband speed, Mbps	2.2
Hindu	<0.1		
Jewish	<0.1		

a Iranian year ending March 20 2018.
b Iranian year ending March 20 2019.

IRELAND

Area, sq km	70,280	Capital	Dublin
Arable as % of total land	6.5	Currency	Euro (€)

People

Population, m	4.8	Life expectancy: men	81.3 yrs
Pop. per sq km	68.8	women	84.3 yrs
Average annual rate of change		Adult literacy	...
in pop. 2015–20, %	1.2	Fertility rate (per woman)	1.8
Pop. aged 0–19, 2020, %	27.2	Urban population, %	63.2
Pop. aged 65 and over, 2020, %	14.6		per 1,000 pop.
No. of men per 100 women	98.6	Crude birth rate	13.0
Human Development Index	94.2	Crude death rate	6.5

The economy

GDP	$382bn	GDP per head	$78,806
GDP	€324bn	GDP per head in purchasing	
Av. ann. growth in real		power parity (USA=100)	132.5
GDP 2013–18	10.5%	Economic freedom index	80.9

Origins of GDP		**Components of GDP**	
	% of total		% of total
Agriculture	1.0	Private consumption	31.0
Industry, of which:	39.3	Public consumption	11.9
manufacturing	34.6	Investment	23.8
Services	59.7	Exports	122.3
		Imports	-89.2

Structure of employment

	% of total		% of labour force
Agriculture	4.6	Unemployed 2019	5.7
Industry	18.7	Av. ann. rate 2009-19	4.9
Services	76.7		

Energy

	m TOE		
Total output	5.3	Net energy imports as %	
Total consumption	16.2	of energy use	67
Consumption per head			
kg oil equivalent	3,396		

Inflation and finance

			% change 2018–19
Consumer price			
inflation 2019	0.9%	Narrow money (M1)	1.8
Av. ann. inflation 2014–19	0.3%	Broad money	5.2
Deposit rate, Dec. 2019	0.02%		

Exchange rates

	end 2019		December 2019
€ per $	0.89	Effective rates	2010 = 100
€ per SDR	1.23	– nominal	96.4
		– real	87.4

Trade

Principal exports		Principal imports	
	$bn fob		*$bn cif*
Chemicals & related products	101.3	Machinery & transport equip.	46.6
Machinery & transport equip.	23.0	Chemicals & related products	23.8
Food, drink and tobacco	14.9	Food, drink and tobacco	10.3
Raw materials	2.3	Mineral fuels & lubricants	6.8
Total incl. others	**166.0**	Total incl. others	**108.3**

Main export destinations		Main origins of imports	
	% of total		*% of total*
United States	27.9	United Kingdom	25.7
Belgium	13.0	United States	18.3
United Kingdom	11.3	Germany	11.9
Germany	7.3	France	11.7
EU28	50.1	EU28	63.8

Balance of payments, reserves and aid, $bn

Visible exports fob	255.3	Overall balance	0.9
Visible imports fob	-121.7	Change in reserves	0.8
Trade balance	133.5	Level of reserves	
Invisibles inflows	310.3	end Dec.	5.2
Invisibles outflows	-399.0	No. months of import cover	0.1
Net transfers	-4.0	Official gold holdings, m oz	0.2
Current account balance	40.9	Aid given	0.9
– as % of GDP	10.7	– as % of GNI	0.3
Capital balance	-45.6		

Health and education

Health spending, % of GDP	7.2	Education spending, % of GDP	3.7
Doctors per 1,000 pop.	3.3	Enrolment, %: primary	101
Hospital beds per 1,000 pop.	2.8	secondary	125
At least basic drinking water,		tertiary	78
% of pop.	97.4		

Society

No. of households, m	1.7	Cost of living, Dec. 2019	
Av. no. per household	2.8	New York = 100	81
Marriages per 1,000 pop.	4.4	Cars per 1,000 pop.	420
Divorces per 1,000 pop.	0.7	Telephone lines per 100 pop.	38.0
Religion, % of pop.		Mobile telephone subscribers	
Christian	92.0	per 100 pop.	103.2
Non-religious	6.2	Internet access, %	84.5
Muslim	1.1	Broadband subs per 100 pop.	29.7
Other	0.4	Broadband speed, Mbps	18.2
Hindu	0.2		
Jewish	<0.1		

ISRAEL

Area, sq km	22,070	Capital	Jerusalem[a]
Arable as % of total land	13.6	Currency	New Shekel (NIS)

People

Population, m	8.4	Life expectancy: men	82.0 yrs
Pop. per sq km	380.6	women	84.9 yrs
Average annual rate of change		Adult literacy	...
in pop. 2015–20, %	1.6	Fertility rate (per woman)	3.0
Pop. aged 0–19, 2020, %	35.5	Urban population, %	92.4
Pop. aged 65 and over, 2020, %	12.4		per 1,000 pop.
No. of men per 100 women	99.1	Crude birth rate	20.4
Human Development Index	90.6	Crude death rate	5.3

The economy

GDP	$371bn	GDP per head	$41,715
GDP	NIS1,331bn	GDP per head in purchasing	
Av. ann. growth in real		power parity (USA=100)	63.6
GDP 2013–18	3.4%	Economic freedom index	74.0

Origins of GDP		**Components of GDP**	
	% of total		% of total
Agriculture	1.3	Private consumption	54.7
Industry, of which:	21.5	Public consumption	23.0
manufacturing	13.2	Investment	21.8
Services	77.2	Exports	29.4
		Imports	-29.0

Structure of employment

	% of total		% of labour force
Agriculture	0.9	Unemployed 2019	4.0
Industry	17.0	Av. ann. rate 2009-19	3.9
Services	82.1		

Energy

	m TOE		
Total output	9.4	Net energy imports as %	
Total consumption	26.7	of energy use	65
Consumption per head			
kg oil equivalent	3,213		

Inflation and finance

			% change 2018–19
Consumer price			
inflation 2019	0.8%	Narrow money (M1)	13.8
Av. ann. inflation 2014–19	0.1%	Broad money	6.6
Deposit rate, Dec. 2019	0.64%		

Exchange rates

	end 2019		December 2019
			2010 = 100
NIS per $	3.46	Effective rates	
NIS per SDR	4.78	– nominal	126.8
NIS per €	3.89	– real	110.4

Trade

Principal exports		Principal imports	
	$bn fob		*$bn cif*
Chemicals & chemical products	13.6	Fuel	9.8
Communications, medical &		Machinery & equipment	8.9
scientific equipment	9.2	Diamonds	5.7
Polished diamonds	6.8	Chemicals	5.2
Electronic components &			
computers	4.9		
Total incl. others	**54.1**	Total incl. others	**75.6**

Main export destinations		Main origins of imports	
	% of total		*% of total*
United States	31.0	United States	12.9
China	8.5	Switzerland	10.3
United Kingdom	8.1	China	9.0
Hong Kong	7.7	United Kingdom	8.1

Balance of payments, reserves and debt, $bn

Visible exports fob	58.9	Change in reserves	2.3
Visible imports fob	-77.1	Level of reserves	
Trade balance	-18.3	end Dec.	115.3
Invisibles inflows	64.6	No. months of import cover	11.3
Invisibles outflows	-44.8	Official gold holdings, m oz	0.0
Net transfers	7.8	Foreign debt	93.8
Current account balance	9.3	– as % of GDP	25.3
– as % of GDP	2.5	– as % of total exports	75.2
Capital balance	4.2	Debt service ratio	11.6
Overall balance	5.5		

Health and education

Health spending, % of GDP	7.4	Education spending, % of GDP	5.8
Doctors per 1,000 pop.	4.6	Enrolment, %: primary	105
Hospital beds per 1,000 pop.	3.1	secondary	105
At least basic drinking water,		tertiary	63
% of pop.	100		

Society

No. of households, m	2.5	Cost of living, Dec. 2019	
Av. no. per household	3.4	New York = 100	97
Marriages per 1,000 pop.	6.2	Cars per 1,000 pop.	328
Divorces per 1,000 pop.	1.7	Telephone lines per 100 pop.	38.2
Religion, % of pop.		Mobile telephone subscribers	
Jewish	75.6	per 100 pop.	127.7
Muslim	18.6	Internet access, %	83.7
Non-religious	3.1	Broadband subs per 100 pop.	28.8
Christian	2.0	Broadband speed, Mbps	7.6
Other	0.6		
Hindu	<0.1		

a Sovereignty over the city is disputed.

ITALY

Area, sq km	301,340	Capital	Rome
Arable as % of total land	22.4	Currency	Euro (€)

People

Population, m	60.6	Life expectancy: men	81.9 yrs
Pop. per sq km	200.6	women	86.0 yrs
Average annual rate of change		Adult literacy	99.2
in pop. 2015–20, %	0.0	Fertility rate (per woman)	1.3
Pop. aged 0–19, 2020, %	17.7	Urban population, %	70.4
Pop. aged 65 and over, 2020, %	23.3		per 1,000 pop.
No. of men per 100 women	94.9	Crude birth rate	7.6
Human Development Index	88.3	Crude death rate	10.9

The economy

GDP	$2,084bn	GDP per head	$34,483
GDP	€1,757bn	GDP per head in purchasing	
Av. ann. growth in real		power parity (USA=100)	66.6
GDP 2013–18	0.9%	Economic freedom index	63.8

Origins of GDP		**Components of GDP**	
	% of total		% of total
Agriculture	2.2	Private consumption	60.3
Industry, of which:	23.9	Public consumption	19.0
manufacturing	16.7	Investment	18.2
Services	73.9	Exports	31.5
		Imports	-29.0

Structure of employment

	% of total		% of labour force
Agriculture	3.7	Unemployed 2019	10.6
Industry	25.9	Av. ann. rate 2009–19	9.9
Services	70.4		

Energy

	m TOE		
Total output	34.0	Net energy imports as %	
Total consumption	171.3	of energy use	80
Consumption per head			
kg oil equivalent	2,886		

Inflation and finance

		% change 2018–19	
Consumer price			
inflation 2019	0.6%	Narrow money (M1)	1.8
Av. ann. inflation 2014–19	0.7%	Broad money	5.2
Deposit rate, Dec. 2019	0.71%		

Exchange rates

	end 2019		December 2019
€ per $	0.89	Effective rates	2010 = 100
€ per SDR	1.23	– nominal	105.2
		– real	93.3

Trade

Principal exports	$bn fob	Principal imports	$bn cif
Machinery & transport equip.	197.7	Machinery & transport equip.	142.8
Chemicals & related products	72.1	Chemicals & related products	79.6
Food, drink and tobacco	45.3	Mineral fuels & lubricants	67.6
Mineral fuels & lubricants	20.4	Food, drink and tobacco	44.0
Total incl. others	**547.3**	Total incl. others	**501.8**

Main export destinations	% of total	Main origins of imports	% of total
Germany	12.5	Germany	16.5
France	10.5	France	8.6
United States	9.1	China	7.3
Spain	5.2	Netherlands	5.3
EU28	56.5	EU28	58.8

Balance of payments, reserves and aid, $bn

Visible exports fob	533.1	Overall balance	3.1
Visible imports fob	-479.5	Change in reserves	1.5
Trade balance	53.6	Level of reserves	
Invisibles inflows	216.0	end Dec.	152.2
Invisibles outflows	-197.4	No. months of import cover	2.7
Net transfers	-20.7	Official gold holdings, m oz	78.8
Current account balance	51.5	Aid given	5.1
– as % of GDP	2.5	– as % of GNI	0.3
Capital balance	-33.1		

Health and education

Health spending, % of GDP	8.8	Education spending, % of GDP	3.8
Doctors per 1,000 pop.	4.0	Enrolment, %: primary	102
Hospital beds per 1,000 pop.	3.4	secondary	101
At least basic drinking water,		tertiary	62
% of pop.	99.4		

Society

No. of households, m	25.9	Cost of living, Dec. 2019	
Av. no. per household	2.3	New York = 100	73
Marriages per 1,000 pop.	3.2	Cars per 1,000 pop.	622
Divorces per 1,000 pop.	1.5	Telephone lines per 100 pop.	33.6
Religion, % of pop.		Mobile telephone subscribers	
Christian	83.3	per 100 pop.	137.5
Non-religious	12.4	Internet access, %	74.4
Muslim	3.7	Broadband subs per 100 pop.	28.1
Other	0.4	Broadband speed, Mbps	15.1
Hindu	0.1		
Jewish	<0.1		

IVORY COAST

Area, sq km	322,460	Capital	Yamoussoukro
Arable as % of total land	9.1	Currency	CFA franc (CFAfr)

People

Population, m	25.1	Life expectancy: men	57.5 yrs
Pop. per sq km	77.8	women	60.1 yrs
Average annual rate of change		Adult literacy	47.2
in pop. 2015–20, %	2.5	Fertility rate (per woman)	4.7
Pop. aged 0–19, 2020, %	52.5	Urban population, %	50.8
Pop. aged 65 and over, 2020, %	2.9		per 1,000 pop.
No. of men per 100 women	101.7	Crude birth rate	35.9
Human Development Index	51.6	Crude death rate	9.4

The economy

GDP	$43bn	GDP per head	$1,716
GDP	CFAfr23,900bn	GDP per head in purchasing	
Av. ann. growth in real		power parity (USA=100)	6.7
GDP 2013–18	8.1%	Economic freedom index	59.7

Origins of GDP[a]		**Components of GDP**	
	% of total		% of total
Agriculture	24.3	Private consumption	65.9
Industry, of which:	27.8	Public consumption	13.8
manufacturing	13.0	Investment	19.8
Services	47.8	Exports	29.8
		Imports	-29.2

Structure of employment

	% of total		% of labour force
Agriculture	40.0	Unemployed 2019	3.2
Industry	13.2	Av. ann. rate 2009-19	3.3
Services	46.8		

Energy

	m TOE		
Total output	5.4	Net energy imports as %	
Total consumption	4.5	of energy use	-19
Consumption per head			
kg oil equivalent	186		

Inflation and finance

Consumer price			% change 2018–19
inflation 2019	0.8%	Narrow money (M1)	22.0
Av. ann. inflation 2014–19	0.8%	Broad money	10.8
Benchmark interest rate, Dec. 2019 4.50%			

Exchange rates

	end 2019		December 2019
			2010 = 100
CFAfr per $	583.90	Effective rates	
CFAfr per SDR	807.44	– nominal	112.7
CFAfr per €	656.07	– real	93.7

Trade

Principal exports		Principal imports	
	$bn fob		*$bn cif*
Cocoa beans & butter	4.6	Fuels & lubricants	2.4
Petroleum products	1.7	Foodstuffs	2.2
Cashew nuts	1.1	Raw materials & intermediate	
Gold	0.8	goods	1.9
		Capital equip.	1.8
Total incl. others	**11.9**	Total incl. others	**11.1**

Main export destinations		Main origins of imports	
	% of total		*% of total*
Netherlands	11.5	China	14.9
United States	9.2	Nigeria	12.3
Vietnam	6.8	France	10.3
Germany	6.4	India	4.5

Balance of payments, reserves and debt, $bn

Visible exports fob	11.9	Change in reserves	0.7
Visible imports fob	-9.5	Level of reserves	
Trade balance	2.5	end Dec.	6.6
Invisibles inflows	1.5	No. months of import cover	5.3
Invisibles outflows	-5.5	Official gold holdings, m oz	0.0
Net transfers	-0.6	Foreign debt	15.7
Current account balance	-2.1	– as % of GDP	36.4
– as % of GDP	-4.8	– as % of total exports	114.7
Capital balance	2.4	Debt service ratio	12.1
Overall balance	0.2		

Health and education

Health spending, % of GDP	4.5	Education spending, % of GDP	4.4
Doctors per 1,000 pop.	...	Enrolment, %: primary	100
Hospital beds per 1,000 pop.	0.4	secondary	51
At least basic drinking water,		tertiary	9
% of pop.	72.9		

Society

No. of households, m	3.7	Cost of living, Dec. 2019	
Av. no. per household	6.8	New York = 100	53
Marriages per 1,000 pop.	...	Cars per 1,000 pop.	18
Divorces per 1,000 pop.	...	Telephone lines per 100 pop.	1.2
Religion, % of pop.	...	Mobile telephone subscribers	
		per 100 pop.	134.9
		Internet access, %	46.8
		Broadband subs per 100 pop.	0.7
		Broadband speed, Mbps	1.7

JAPAN

Area, sq km	377,930	Capital	Tokyo
Arable as % of total land	11.5	Currency	Yen (¥)

People

Population, m	127.2	Life expectancy: men	81.9 yrs
Pop. per sq km	336.6	women	88.1 yrs
Average annual rate of change		Adult literacy	...
in pop. 2015–20, %	-0.2	Fertility rate (per woman)	1.4
Pop. aged 0–19, 2020, %	17.0	Urban population, %	91.6
Pop. aged 65 and over, 2020, %	28.4		per 1,000 pop.
No. of men per 100 women	95.4	Crude birth rate	7.5
Human Development Index	91.5	Crude death rate	11.5

The economy

GDP	$4,971bn	GDP per head	$39,290
GDP	Yen 548trn	GDP per head in purchasing	
Av. ann. growth in real		power parity (USA=100)	68.2
GDP 2013–18	1.0%	Economic freedom index	73.3

Origins of GDP		**Components of GDP**	
	% of total		% of total
Agriculture	1.2	Private consumption	55.6
Industry, of which:	26.5	Public consumption	19.7
manufacturing	20.7	Investment	24.4
Services	71.6	Exports	18.4
		Imports	-18.2

Structure of employment

	% of total		% of labour force
Agriculture	3.4	Unemployed 2019	2.4
Industry	24.3	Av. ann. rate 2009–19	2.3
Services	72.3		

Energy

	m TOE		
Total output	55.8	Net energy imports as %	
Total consumption	494.4	of energy use	89
Consumption per head			
kg oil equivalent	3,878		

Inflation and finance

			% change 2018–19
Consumer price			
inflation 2019	0.5%	Narrow money (M1)	3.1
Av. ann. inflation 2014–19	0.5%	Broad money	1.9
Central bank policy rate, Dec. 2019	-0.10%		

Exchange rates

	end 2019		December 2019
¥ per $	109.12	Effective rates	2010 = 100
¥ per SDR	150.89	– nominal	88.7
¥ per €	122.61	– real	76.0

Trade

Principal exports		Principal imports	
	$bn fob		*$bn cif*
Capital equipment	372.6	Industrial supplies	346.7
Industrial supplies	174.4	Capital equipment	206.8
Consumer durable goods	121.4	Food & direct consumer goods	64.4
Consumer non-durable goods	8.3	Consumer durable goods	58.5
Total incl. others	**737.9**	Total incl. others	**749.0**

Main export destinations		Main origins of imports	
	% of total		*% of total*
China	19.5	China	23.2
United States	19.1	United States	11.2
South Korea	7.1	Australia	6.1
Taiwan	5.7	Saudi Arabia	4.5

Balance of payments, reserves and aid, $bn

Visible exports fob	735.6	Overall balance	23.9
Visible imports fob	-725.0	Change in reserves	6.4
Trade balance	10.6	Level of reserves	
Invisibles inflows	495.9	end Dec.	1,270.4
Invisibles outflows	-312.2	No. months of import cover	14.7
Net transfers	-18.2	Official gold holdings, m oz	24.6
Current account balance	176.1	Aid given	10.1
– as % of GDP	3.5	– as % of GNI	0.2
Capital balance	-159.9		

Health and education

Health spending, % of GDP	10.9	Education spending, % of GDP	3.2
Doctors per 1,000 pop.	2.4	Enrolment, %: primary	...
Hospital beds per 1,000 pop.	13.4	secondary	...
At least basic drinking water,		tertiary	...
% of pop.	99.0		

Society

No. of households, m	54.2	Cost of living, Dec. 2019	
Av. no. per household	2.3	New York = 100	96
Marriages per 1,000 pop.	4.8	Cars per 1,000 pop.	485
Divorces per 1,000 pop.	1.7	Telephone lines per 100 pop.	49.9
Religion, % of pop.		Mobile telephone subscribers	
Non-religious	57.0	per 100 pop.	141.4
Buddhist	36.2	Internet access, %	91.3
Other	5.0	Broadband subs per 100 pop.	32.6
Christian	1.6	Broadband speed, Mbps	28.9
Muslim	0.2		
Jewish	<0.1		

KENYA

Area, sq km	580,370	Capital	Nairobi
Arable as % of total land	10.2	Currency	Kenyan shilling (KSh)

People

Population, m	51.4	Life expectancy: men	65.0 yrs
Pop. per sq km	86.8	women	69.9 yrs
Average annual rate of change		Adult literacy	81.5
in pop. 2015–20, %	2.3	Fertility rate (per woman)	3.5
Pop. aged 0–19, 2020, %	49.8	Urban population, %	27.0
Pop. aged 65 and over, 2020, %	2.5		per 1,000 pop.
No. of men per 100 women	98.8	Crude birth rate	28.9
Human Development Index	57.9	Crude death rate	5.3

The economy

GDP	$88bn	GDP per head	$1,711
GDP	KSh8,905bn	GDP per head in purchasing	
Av. ann. growth in real		power parity (USA=100)	5.5
GDP 2013–18	5.6%	Economic freedom index	55.3

Origins of GDP		**Components of GDP**	
	% of total		% of total
Agriculture	37.4	Private consumption	81.8
Industry, of which:	17.9	Public consumption	12.9
manufacturing	8.5	Investment	18.4
Services	44.6	Exports	13.2
		Imports	-23.0

Structure of employment

	% of total		% of labour force
Agriculture	54.4	Unemployed 2019	2.6
Industry	7.3	Av. ann. rate 2009–19	2.6
Services	38.2		

Energy

	m TOE		
Total output	1.9	Net energy imports as %	
Total consumption	8.1	of energy use	77
Consumption per head			
kg oil equivalent	164		

Inflation and finance

			% change 2018–19
Consumer price			
inflation 2019	5.2%	Narrow money (M1)	-4.0
Av. ann. inflation 2014–19	6.1%	Broad money	5.6
Deposit rate, Sep. 2019	7.57%		

Exchange rates

	end 2019		December 2019
			2010 = 100
KSh per $	101.34	Effective rates	
KSh per SDR	140.13	– nominal	...
KSh per €	113.87	– real	...

Trade

Principal exports

	$bn fob
Tea	1.4
Horticultural products	0.8
Coffee	0.2
Total incl. others	**5.9**

Principal imports

	$bn cif
Industrial supplies	6.0
Machinery & other capital equip.	2.9
Transport equipment	1.9
Food & beverages	1.7
Total incl. others	**16.3**

Main export destinations

	% of total
Uganda	10.1
Pakistan	9.7
United States	7.7
Netherlands	7.6

Main origins of imports

	% of total
China	21.0
India	10.5
Saudi Arabia	9.8
United Arab Emirates	8.4

Balance of payments, reserves and debt, $bn

Visible exports fob	6.1	Change in reserves	0.8
Visible imports fob	-16.3	Level of reserves	
Trade balance	-10.2	end Dec.	8.2
Invisibles inflows	6.2	No. months of import cover	4.5
Invisibles outflows	-5.3	Official gold holdings, m oz	0.0
Net transfers	5.0	Foreign debt	31.5
Current account balance	-4.4	– as % of GDP	35.8
– as % of GDP	-5.0	– as % of total exports	218.0
Capital balance	6.8	Debt service ratio	18.6
Overall balance	1.0		

Health and education

Health spending, % of GDP	4.8	Education spending, % of GDP	5.3
Doctors per 1,000 pop.	0.2	Enrolment, %: primary	103
Hospital beds per 1,000 pop.	1.4	secondary	…
At least basic drinking water, % of pop.	58.9	tertiary	11

Society

No. of households, m	11.2	Cost of living, Dec. 2019	
Av. no. per household	4.6	New York = 100	66
Marriages per 1,000 pop.	…	Cars per 1,000 pop.	18
Divorces per 1,000 pop.	…	Telephone lines per 100 pop.	0.1
Religion, % of pop.		Mobile telephone subscribers	
Christian	84.8	per 100 pop.	96.3
Muslim	9.7	Internet access, %	17.8
Other	3.0	Broadband subs per 100 pop.	0.7
Non-religious	2.5	Broadband speed, Mbps	10.1
Hindu	0.1		
Jewish	<0.1		

MALAYSIA

Area, sq km	330,345	Capital	Kuala Lumpur
Arable as % of total land	2.7	Currency	Malaysian dollar/ringgit (M$)

People

Population, m	31.5	Life expectancy: men	74.7 yrs
Pop. per sq km	95.3	women	78.8 yrs
Average annual rate of change		Adult literacy	94.9
in pop. 2015–20, %	1.3	Fertility rate (per woman)	2.0
Pop. aged 0–19, 2020, %	31.7	Urban population, %	76.0
Pop. aged 65 and over, 2020, %	7.2		per 1,000 pop.
No. of men per 100 women	105.7	Crude birth rate	16.8
Human Development Index	80.4	Crude death rate	5.5

The economy

GDP	$359bn	GDP per head	$11,373
GDP	M$1,447bn	GDP per head in purchasing	
Av. ann. growth in real		power parity (USA=100)	50.6
GDP 2013–18	5.2%	Economic freedom index	74.7

Origins of GDP		**Components of GDP**	
	% of total		% of total
Agriculture	7.5	Private consumption	57.4
Industry, of which:	38.3	Public consumption	12.0
manufacturing	21.6	Investment	23.6
Services	54.2	Exports	68.8
		Imports	-61.7

Structure of employment

	% of total		% of labour force
Agriculture	10.4	Unemployed 2019	3.3
Industry	27.0	Av. ann. rate 2009-19	3.3
Services	62.6		

Energy

	m TOE		
Total output	109.3	Net energy imports as %	
Total consumption	88.6	of energy use	-23
Consumption per head			
kg oil equivalent	2,803		

Inflation and finance

			% change 2018–19
Consumer price			
inflation 2019	0.7%	Narrow money (M1)	2.6
Av. ann. inflation 2014–19	1.9%	Broad money	2.7
Deposit rate, Dec. 2019	2.9%		

Exchange rates

	end 2019		December 2019
M$ per $	4.09	Effective rates	2010 = 100
M$ per SDR	5.66	– nominal	87.0
M$ per €	4.60	– real	86.7

Trade

Principal exports	$bn fob	Principal imports	$bn cif
Machinery & transport equip.	108.9	Machinery & transport equip.	94.3
Mineral fuels	38.5	Mineral fuels	31.4
Manufactured goods	22.8	Manufactured goods	25.5
Chemicals	21.0	Chemicals	23.5
Total incl. others	**247.4**	Total incl. others	**217.5**

Main export destinations	% of total	Main origins of imports	% of total
China	13.9	China	20.0
Singapore	13.9	Singapore	11.7
United States	9.1	United States	7.4
Hong Kong	7.5	Japan	7.3

Balance of payments, reserves and debt, $bn

Visible exports fob	206.3	Change in reserves	-1.0
Visible imports fob	-176.8	Level of reserves	
Trade balance	29.6	end Dec.	101.4
Invisibles inflows	53.4	No. months of import cover	4.9
Invisibles outflows	-70.6	Official gold holdings, m oz	1.3
Net transfers	-4.8	Foreign debt	223.5
Current account balance	7.6	– as % of GDP	62.3
– as % of GDP	2.1	– as % of total exports	85.5
Capital balance	4.8	Debt service ratio	4.1
Overall balance	2.1		

Health and education

Health spending, % of GDP	3.9	Education spending, % of GDP	4.5
Doctors per 1,000 pop.	1.5	Enrolment, %: primary	105
Hospital beds per 1,000 pop.	1.9	secondary	82
At least basic drinking water,		tertiary	45
% of pop.	96.7		

Society

No. of households, m	7.7	Cost of living, Dec. 2019	
Av. no. per household	4.1	New York = 100	57
Marriages per 1,000 pop.	5.9	Cars per 1,000 pop.	409
Divorces per 1,000 pop.	...	Telephone lines per 100 pop.	20.4
Religion, % of pop.		Mobile telephone subscribers	
Muslim	63.7	per 100 pop.	134.5
Buddhist	17.7	Internet access, %	81.2
Christian	9.4	Broadband subs per 100 pop.	8.6
Hindu	6.0	Broadband speed, Mbps	13.3
Other	2.5		
Non-religious	0.7		

MEXICO

Area, sq km	1,964,375	Capital	Mexico City
Arable as % of total land	11.6	Currency	Mexican peso (PS)

People

Population, m	126.2	Life expectancy: men	72.6 yrs
Pop. per sq km	64.2	women	78.2 yrs
Average annual rate of change		Adult literacy	95.4
in pop. 2015–20, %	1.1	Fertility rate (per woman)	2.1
Pop. aged 0–19, 2020, %	34.5	Urban population, %	80.2
Pop. aged 65 and over, 2020, %	7.6		per 1,000 pop.
No. of men per 100 women	95.8	Crude birth rate	17.7
Human Development Index	76.7	Crude death rate	6.3

The economy

GDP	$1,221bn	GDP per head	$9,673
GDP	PS23,518bn	GDP per head in purchasing	
Av. ann. growth in real		power parity (USA=100)	31.6
GDP 2013–18	2.7%	Economic freedom index	66.0

Origins of GDP		Components of GDP	
	% of total		% of total
Agriculture	3.6	Private consumption	64.8
Industry, of which:	32.8	Public consumption	11.7
manufacturing	18.3	Investment	22.7
Services	63.7	Exports	39.3
		Imports	-41.2

Structure of employment

	% of total		% of labour force
Agriculture	12.6	Unemployed 2019	3.3
Industry	26.1	Av. ann. rate 2009–19	3.4
Services	61.2		

Energy

	m TOE		
Total output	168.3	Net energy imports as %	
Total consumption	199.5	of energy use	16
Consumption per head			
kg oil equivalent	1,544		

Inflation and finance

			% change 2018–19
Consumer price			
inflation 2019	3.6%	Narrow money (M1)	4.1
Av. ann. inflation 2014–19	4.0%	Broad money	6.8
Deposit rate, Dec. 2019	3.11%		

Exchange rates

	end 2019		December 2019
			2010 = 100
PS per $	18.85	Effective rates	
PS per SDR	26.06	– nominal	71.5
PS per €	21.18	– real	84.6

Trade

Principal exports	$bn fob	Principal imports	$bn cif
Manufactured goods	397.3	Intermediate goods	355.3
Crude oil & products	30.1	Consumer goods	63.1
Agricultural products	16.5	Capital goods	45.9
Mining products	6.2		
Total incl. others	**450.7**	Total	**464.3**

Main export destinations	% of total	Main origins of imports	% of total
United States	79.5	United States	49.3
Canada	3.1	China	19.1
China	1.6	Japan	4.2
Germany	1.6	Germany	4.1

Balance of payments, reserves and debt, $bn

Visible exports fob	451.1	Change in reserves	0.9
Visible imports fob	-464.8	Level of reserves	
Trade balance	-13.8	end Dec.	176.4
Invisibles inflows	41.7	No. months of import cover	3.9
Invisibles outflows	-83.8	Official gold holdings, m oz	3.9
Net transfers	32.9	Foreign debt	453.0
Current account balance	-23.0	– as % of GDP	37.1
– as % of GDP	-1.9	– as % of total exports	86.1
Capital balance	33.4	Debt service ratio	11.1
Overall balance	0.5		

Health and education

Health spending, % of GDP	5.5	Education spending, % of GDP	4.9
Doctors per 1,000 pop.	2.4	Enrolment, %: primary	106
Hospital beds per 1,000 pop.	1.5	secondary	104
At least basic drinking water,		tertiary	40
% of pop.	99.3		

Society

No. of households, m	33.8	Cost of living, Dec. 2019	
Av. no. per household	3.7	New York = 100	70
Marriages per 1,000 pop.	4.0	Cars per 1,000 pop.	229
Divorces per 1,000 pop.	1.2	Telephone lines per 100 pop.	17.2
Religion, % of pop.		Mobile telephone subscribers	
Christian	95.1	per 100 pop.	95.2
Non-religious	4.7	Internet access, %	65.8
Hindu	<0.1	Broadband subs per 100 pop.	14.6
Jewish	<0.1	Broadband speed, Mbps	5.7
Muslim	<0.1		
Other	<0.1		

MOROCCO

Area, sq km	447,400	Capital	Rabat
Arable as % of total land	18.2	Currency	Dirham (Dh)

People

Population, m	36.0	Life expectancy: men	76.2 yrs
Pop. per sq km	80.5	women	78.7 yrs
Average annual rate of change		Adult literacy	73.8
in pop. 2015–20, %	1.3	Fertility rate (per woman)	2.4
Pop. aged 0–19, 2020, %	34.8	Urban population, %	62.5
Pop. aged 65 and over, 2020, %	7.6		per 1,000 pop.
No. of men per 100 women	98.5	Crude birth rate	19.1
Human Development Index	67.6	Crude death rate	5.1

The economy

GDP	$118bn	GDP per head	$3,222
GDP	Dh1,113bn	GDP per head in purchasing	
Av. ann. growth in real		power parity (USA=100)	13.7
GDP 2013–18	3.1%	Economic freedom index	63.3

Origins of GDP

	% of total
Agriculture	14.0
Industry, of which:	29.6
manufacturing	17.8
Services	56.7

Components of GDP

	% of total
Private consumption	58.0
Public consumption	19.0
Investment	33.5
Exports	38.7
Imports	-49.3

Structure of employment

	% of total		% of labour force
Agriculture	34.7	Unemployed 2019	9.1
Industry	21.7	Av. ann. rate 2009-19	9.0
Services	43.6		

Energy

	m TOE		
Total output	1.2	Net energy imports as %	
Total consumption	20.7	of energy use	94
Consumption per head			
kg oil equivalent	578		

Inflation and finance

			% change 2018–19
Consumer price			
inflation 2019	0.0%	Narrow money (M1)	5.8
Av. ann. inflation 2014–19	1.1%	Broad money	3.3
Deposit rate, Dec. 2019	2.98%		

Exchange rates

	end 2019		December 2019
Dh per $	9.59	Effective rates	2010 = 100
Dh per SDR	13.27	– nominal	110.0
Dh per €	10.78	– real	100.1

Trade

Principal exports		Principal imports	
	$bn fob		*$bn cif*
Electric cables & wires	3.2	Capital goods	12.8
Fertilisers & chemicals	2.5	Consumer goods	11.6
Phosphoric acid	1.5	Semi-finished goods	10.6
Finished clothes	0.5	Fuel & lubricants	8.8
Total incl. others	**29.5**	Total incl. others	**51.5**

Main export destinations		Main origins of imports	
	% of total		*% of total*
Spain	22.6	Spain	15.7
France	22.1	France	11.9
Brazil	6.1	China	9.9
United States	5.5	United States	8.0

Balance of payments, reserves and debt, $bn

Visible exports fob	24.6	Change in reserves	-1.7
Visible imports fob	-44.9	Level of reserves	
Trade balance	-20.3	end Dec.	24.5
Invisibles inflows	19.4	No. months of import cover	5.0
Invisibles outflows	-13.3	Official gold holdings, m oz	0.7
Net transfers	8.0	Foreign debt	49.0
Current account balance	-6.2	– as % of GDP	41.6
– as % of GDP	-5.3	– as % of total exports	96.7
Capital balance	3.9	Debt service ratio	7.6
Overall balance	-1.0		

Health and education

Health spending, % of GDP	5.2	Education spending, % of GDP	...
Doctors per 1,000 pop.	0.7	Enrolment, %: primary	114
Hospital beds per 1,000 pop.	1.1	secondary	80
At least basic drinking water,		tertiary	36
% of pop.	86.8		

Society

No. of households, m	7.9	Cost of living, Dec. 2019	
Av. no. per household	5.2	New York = 100	52
Marriages per 1,000 pop.	...	Cars per 1,000 pop.	76
Divorces per 1,000 pop.	...	Telephone lines per 100 pop.	6.1
Religion, % of pop.		Mobile telephone subscribers	
Muslim	99.9	per 100 pop.	124.2
Christian	<0.1	Internet access, %	64.8
Hindu	<0.1	Broadband subs per 100 pop.	4.3
Jewish	<0.1	Broadband speed, Mbps	4.0
Non-religious	<0.1		
Other	<0.1		

NETHERLANDS

Area, sq km³	41,540	Capital	Amsterdam
Arable as % of total land	30.5	Currency	Euro (€)

People

Population, m	17.1	Life expectancy: men	81.2 yrs
Pop. per sq km	457.8	women	84.4 yrs
Average annual rate of change		Adult literacy	...
in pop. 2015–20, %	0.2	Fertility rate (per woman)	1.7
Pop. aged 0–19, 2020, %	21.6	Urban population, %	91.5
Pop. aged 65 and over, 2020, %	20.0		per 1,000 pop.
No. of men per 100 women	99.3	Crude birth rate	10.1
Human Development Index	93.3	Crude death rate	9.2

The economy

GDP	$914bn	GDP per head	$53,024
GDP	€774bn	GDP per head in purchasing	
Av. ann. growth in real		power parity (USA=100)	89.7
GDP 2013–18	2.2%	Economic freedom index	77.0

Origins of GDP		**Components of GDP**	
	% of total		% of total
Agriculture	1.8	Private consumption	44.1
Industry, of which:	20.0	Public consumption	24.2
manufacturing	12.4	Investment	20.7
Services	78.1	Exports	84.3
		Imports	-73.3

Structure of employment

	% of total		% of labour force
Agriculture	2.0	Unemployed 2019	3.8
Industry	16.0	Av. ann. rate 2009-19	3.2
Services	82.0		

Energy

	m TOE		
Total output	42.6	Net energy imports as %	
Total consumption	98.3	of energy use	57
Consumption per head			
kg oil equivalent	5,768		

Inflation and finance

			% change 2018–19
Consumer price			
inflation 2019	2.7%	Narrow money (M1)	1.8
Av. ann. inflation 2014–19	1.2%	Broad money	5.2
Deposit rate, Dec. 2019	1.15%		

Exchange rates

	end 2019		December 2019
			2010 = 100
€ per $	0.89	Effective rates	
€ per SDR	1.23	– nominal	100.9
		– real	98.9

Trade

Principal exports

	$bn fob
Machinery & transport equip.	234.5
Chemicals & related products	123.7
Mineral fuels & lubricants	97.7
Food, drink & tobacco	87.3
Total incl. others	**587.9**

Principal imports

	$bn cif
Machinery & transport equip.	217.2
Mineral fuels & lubricants	110.5
Chemicals & related products	81.8
Food, drink & tobacco	59.2
Total incl. others	**521.0**

Main export destinations

	% of total
Germany	23.9
Belgium	13.7
France	10.5
United Kingdom	10.1
EU28	74.2

Main origins of imports

	% of total
China	19.3
Germany	18.5
Belgium	10.1
United States	8.9
EU28	45.6

Balance of payments, reserves and aid, $bn

Visible exports fob	574.6	Overall balance	0.3
Visible imports fob	-487.2	Change in reserves	0.1
Trade balance	87.3	Level of reserves	
Invisibles inflows	528.8	end Dec.	38.4
Invisibles outflows	-509.8	No. months of import cover	0.5
Net transfers	-7.2	Official gold holdings, m oz	19.7
Current account balance	99.1	Aid given	5.6
– as % of GDP	10.8	– as % of GNI	0.6
Capital balance	-105.3		

Health and education

Health spending, % of GDP	10.1	Education spending, % of GDP	5.5
Doctors per 1,000 pop.	3.6	Enrolment, %: primary	104
Hospital beds per 1,000 pop.	4.7	secondary	136
At least basic drinking water,		tertiary	85
% of pop.	100		

Society

No. of households, m	7.8	Cost of living, Dec. 2019	
Av. no. per household	2.2	New York = 100	73
Marriages per 1,000 pop.	3.7	Cars per 1,000 pop.	500
Divorces per 1,000 pop.	1.9	Telephone lines per 100 pop.	34.6
Religion, % of pop.		Mobile telephone subscribers	
Christian	50.6	per 100 pop.	123.7
Non-religious	42.1	Internet access, %	94.7
Muslim	6.0	Broadband subs per 100 pop.	43.4
Other	0.6	Broadband speed, Mbps	36.0
Hindu	0.5		
Jewish	0.2		

a Includes water.

NEW ZEALAND

Area, sq km	267,710	Capital	Wellington
Arable as % of total land	2.2	Currency	New Zealand dollar (NZ$)

People

Population, m	4.7	Life expectancy: men	81.2 yrs
Pop. per sq km	17.5	women	84.4 yrs
Average annual rate of change		Adult literacy	...
in pop. 2015–20, %	0.9	Fertility rate (per woman)	1.9
Pop. aged 0–19, 2020, %	25.7	Urban population, %	86.5
Pop. aged 65 and over, 2020, %	16.4		per 1,000 pop.
No. of men per 100 women	96.7	Crude birth rate	12.6
Human Development Index	92.1	Crude death rate	7.2

The economy

GDP	$205bn	GDP per head	$41,945
GDP	NZ$293bn	GDP per head in purchasing	
Av. ann. growth in real		power parity (USA=100)	65.3
GDP 2013–18	3.4%	Economic freedom index	84.1

Origins of GDP		**Components of GDP**	
	% of total		% of total
Agriculture	5.8	Private consumption	58.1
Industry, of which:	21.7	Public consumption	18.2
manufacturing	10.7	Investment	23.9
Services	72.5	Exports	28.1
		Imports	-28.3

Structure of employment

	% of total		% of labour force
Agriculture	5.7	Unemployed 2019	4.3
Industry	19.6	Av. ann. rate 2009–19	4.1
Services	74.8		

Energy

	m TOE		
Total output	16.4	Net energy imports as %	
Total consumption	23.1	of energy use	29
Consumption per head			
kg oil equivalent	4,919		

Inflation and finance

			% change 2018–19
Consumer price			
inflation 2019	1.6%	Narrow money (M1)	2.5
Av. ann. inflation 2014–19	1.2%	Broad money	4.7
Deposit rate, Dec. 2019	2.67%		

Exchange rates

	end 2019		December 2019
			2010 = 100
NZ$ per $	1.48	Effective rates	
NZ$ per SDR	2.05	– nominal	105.5
NZ$ per €	1.66	– real	101.7

Trade

Principal exports		Principal imports	
	$bn fob		*$bn cif*
Dairy produce	9.9	Machinery & electrical equip.	9.6
Meat	5.1	Transport equipment	7.4
Forestry products	3.6	Mineral fuels	5.4
Fruit	2.2	Textiles	1.9
Total incl. others	**39.8**	Total incl. others	**43.8**

Main export destinations		Main origins of imports	
	% of total		*% of total*
China	24.1	China	19.7
Australia	16.0	Australia	11.5
United States	9.8	United States	10.2
Japan	6.2	Japan	6.9

Balance of payments, reserves and aid, $bn

Visible exports fob	39.8	Overall balance	-2.1
Visible imports fob	-43.3	Change in reserves	-3.0
Trade balance	-3.5	Level of reserves	
Invisibles inflows	23.8	end Dec.	17.7
Invisibles outflows	-27.8	No. months of import cover	3.0
Net transfers	-0.2	Official gold holdings, m oz	0.0
Current account balance	-7.7	Aid given	0.6
– as % of GDP	-3.8	– as % of GNI	0.2
Capital balance	-0.8		

Health and education

Health spending, % of GDP	9.2	Education spending, % of GDP	6.4
Doctors per 1,000 pop.	3.6	Enrolment, %: primary	100
Hospital beds per 1,000 pop.	2.8	secondary	115
At least basic drinking water,		tertiary	82
% of pop.	100		

Society

No. of households, m	1.7	Cost of living, Dec. 2019	
Av. no. per household	2.8	New York = 100	74
Marriages per 1,000 pop.	4.3	Cars per 1,000 pop.	685
Divorces per 1,000 pop.	1.5	Telephone lines per 100 pop.	37.1
Religion, % of pop.		Mobile telephone subscribers	
Christian	57.0	per 100 pop.	134.9
Non-religious	36.6	Internet access, %	90.8
Other	2.8	Broadband subs per 100 pop.	34.7
Hindu	2.1	Broadband speed, Mbps	23.8
Muslim	1.2		
Jewish	0.2		

NIGERIA

Area, sq km	923,768	Capital	Abuja
Arable as % of total land	37.3	Currency	Naira (N)

People

Population, m	195.9	Life expectancy: men	54.8 yrs
Pop. per sq km	212.1	women	56.8 yrs
Average annual rate of change		Adult literacy	62.0
in pop. 2015–20, %	2.6	Fertility rate (per woman)	5.4
Pop. aged 0–19, 2020, %	54.1	Urban population, %	50.3
Pop. aged 65 and over, 2020, %	2.7		per 1,000 pop.
No. of men per 100 women	102.8	Crude birth rate	38.1
Human Development Index	53.4	Crude death rate	11.0

The economy

GDP	$397bn	GDP per head	$2,028
GDP	N129trn	GDP per head in purchasing	
Av. ann. growth in real		power parity (USA=100)	9.5
GDP 2013–18	2.0%	Economic freedom index	57.2

Origins of GDP		**Components of GDP**	
	% of total		% of total
Agriculture	21.4	Private consumption	76.6
Industry, of which:	26.0	Public consumption	5.6
manufacturing	9.7	Investment	19.8
Services	52.6	Exports	15.5
		Imports	-17.5

Structure of employment

	% of total		% of labour force
Agriculture	35.1	Unemployed 2019	8.2
Industry	12.2	Av. ann. rate 2009–19	8.1
Services	52.7		

Energy

	m TOE		
Total output	150.1	Net energy imports as %	
Total consumption	38.8	of energy use	-287
Consumption per head			
kg oil equivalent	203		

Inflation and finance

			% change 2018–19
Consumer price			
inflation 2019	11.4%	Narrow money (M1)	20.7
Av. ann. inflation 2014–19	12.9%	Broad money	6.2
Deposit rate, Dec. 2019	7.95%		

Exchange rates

	end 2019		December 2019
N per $	307.00	Effective rates	2010 = 100
N per SDR	424.53	– nominal	58.1
N per €	344.94	– real	128.7

Trade

Principal exports		Principal imports	
	$bn fob		*$bn cif*
Crude oil	51.4	Machinery & transport equip.	15.4
Gas	7.1	Mineral fuels	12.8
Food, drink & tobacco	0.6	Chemicals	4.9
Vegetable products	0.6	Food & live animals	4.2
Total incl. others	**62.4**	Total incl. others	**43.0**

Main export destinations		Main origins of imports	
	% of total		*% of total*
India	32.0	China	19.0
United States	7.6	United States	8.7
Spain	7.0	Belgium	8.3
France	6.8	United Kingdom	4.5

Balance of payments, reserves and debt, $bn

Visible exports fob	63.1	Change in reserves	3.2
Visible imports fob	-40.8	Level of reserves	
Trade balance	22.3	end Dec.	43.7
Invisibles inflows	6.9	No. months of import cover	5.9
Invisibles outflows	-48.0	Official gold holdings, m oz	0.7
Net transfers	24.1	Foreign debt	47.0
Current account balance	5.3	– as % of GDP	11.2
– as % of GDP	1.3	– as % of total exports	49.7
Capital balance	4.2	Debt service ratio	6.1
Overall balance	3.3		

Health and education

Health spending, % of GDP	3.8	Education spending, % of GDP	...
Doctors per 1,000 pop.	0.4	Enrolment, %: primary	85
Hospital beds per 1,000 pop.	0.5	secondary	42
At least basic drinking water,		tertiary	...
% of pop.	71.4		

Society

No. of households, m	40.6	Cost of living, Dec. 2019	
Av. no. per household	4.8	New York = 100	42
Marriages per 1,000 pop.	...	Cars per 1,000 pop.	16
Divorces per 1,000 pop.	...	Telephone lines per 100 pop.	0.1
Religion, % of pop.		Mobile telephone subscribers	
Christian	49.3	per 100 pop.	88.2
Muslim	48.8	Internet access, %	42
Other	1.4	Broadband subs per 100 pop.	0.1
Non-religious	0.4	Broadband speed, Mbps	1.9
Hindu	<0.1		
Jewish	<0.1		

NORWAY

Area, sq km	385,178	Capital	Oslo
Arable as % of total land	2.2	Currency	Norwegian krone (Nkr)

People

Population, m	5.3	Life expectancy: men	81.1 yrs
Pop. per sq km	16.4	women	84.8 yrs
Average annual rate of change		Adult literacy	...
in pop. 2015–20, %	0.8	Fertility rate (per woman)	1.7
Pop. aged 0–19, 2020, %	23.2	Urban population, %	82.2
Pop. aged 65 and over, 2020, %	17.5		per 1,000 pop.
No. of men per 100 women	102.2	Crude birth rate	11.1
Human Development Index	95.4	Crude death rate	7.9

The economy

GDP	$434bn	GDP per head	$81,697
GDP	Nkr3,531bn	GDP per head in purchasing	
Av. ann. growth in real		power parity (USA=100)	104.3
GDP 2013–18	1.7%	Economic freedom index	73.4

Origins of GDP

Components of GDP

	% of total		% of total
Agriculture	2.2	Private consumption	43.4
Industry, of which:	33.6	Public consumption	23.4
manufacturing	6.6	Investment	27.3
Services	64.1	Exports	38.4
		Imports	-32.6

Structure of employment

	% of total		% of labour force
Agriculture	2.1	Unemployed 2019	3.8
Industry	19.3	Av. ann. rate 2009–19	3.3
Services	78.7		

Energy

	m TOE		
Total output	247.6	Net energy imports as %	
Total consumption	47.7	of energy use	-419
Consumption per head			
kg oil equivalent	8,994		

Inflation and finance

			% change 2018–19
Consumer price			
inflation 2019	2.2%	Narrow money (M1)	17.5
Av. ann. inflation 2014–19	2.5%	Broad money	4.0
Central bank policy rate, Dec. 2019	1.50%		

Exchange rates

	end 2019		December 2019
			2010 = 100
Nkr per $	8.78	Effective rates	
Nkr per SDR	12.14	– nominal	79.8
Nkr per €	9.87	– real	81.5

Trade

Principal exports		Principal imports	
	$bn fob		*$bn cif*
Mineral fuels & lubricants	76.3	Machinery & transport equip.	33.5
Food & beverages	12.7	Manufactured goods	12.6
Manufactured goods	10.5	Miscellaneous manufactured	
Machinery & transport equip.	10.3	goods	13.1
		Chemicals & mineral products	8.9
Total incl. others	**122.6**	Total incl. others	**87.4**

Main export destinations		Main origins of imports	
	% of total		*% of total*
United Kingdom	21.3	Sweden	11.7
Germany	15.8	Germany	11.5
Netherlands	11.0	China	9.4
Sweden	6.7	United Kingdom	7.4
EU28	82.0	EU28	63.6

Balance of payments, reserves and aid, $bn

Visible exports fob	121.9	Overall balance	-0.3
Visible imports fob	-88.3	Change in reserves	-2.8
Trade balance	33.6	Level of reserves	
Invisibles inflows	90.6	end Dec.	63.1
Invisibles outflows	-86.4	No. months of import cover	4.3
Net transfers	-6.6	Official gold holdings, m oz	0.0
Current account balance	31.1	Aid given	4.3
– as % of GDP	7.2	– as % of GNI	1.0
Capital balance	-31.0		

Health and education

Health spending, % of GDP	10.4	Education spending, % of GDP	8.0
Doctors per 1,000 pop.	2.9	Enrolment, %: primary	100
Hospital beds per 1,000 pop.	3.9	secondary	117
At least basic drinking water,		tertiary	82
% of pop.	100		

Society

No. of households, m	2.4	Cost of living, Dec. 2019	
Av. no. per household	2.2	New York = 100	90
Marriages per 1,000 pop.	4.1	Cars per 1,000 pop.	503
Divorces per 1,000 pop.	2.0	Telephone lines per 100 pop.	10.5
Religion, % of pop.		Mobile telephone subscribers	
Christian	84.7	per 100 pop.	107.2
Non-religious	10.1	Internet access, %	96.5
Muslim	3.7	Broadband subs per 100 pop.	41.3
Other	0.9	Broadband speed, Mbps	40.1
Hindu	0.5		
Jewish	<0.1		

PAKISTAN

Area, sq km	796,095	Capital	Islamabad
Arable as % of total land	40.3	Currency	Pakistan rupee (PRs)

People

Population, m	212.2	Life expectancy: men	66.8 yrs
Pop. per sq km	266.6	women	68.9 yrs
Average annual rate of change		Adult literacy	59.1
in pop. 2015–20, %	2.0	Fertility rate (per woman)	3.6
Pop. aged 0–19, 2020, %	44.8	Urban population, %	36.7
Pop. aged 65 and over, 2020, %	4.3		per 1,000 pop.
No. of men per 100 women	106.0	Crude birth rate	28.5
Human Development Index	56.0	Crude death rate	6.7

The economy

GDP	$315bn	GDP per head	$1,482
GDP	PRs34,619bn	GDP per head in purchasing	
Av. ann. growth in real		power parity (USA=100)	8.9
GDP 2013–18	5.3%	Economic freedom index	54.8

Origins of GDP		**Components of GDP**	
	% of total		% of total
Agriculture	24.4	Private consumption	82.5
Industry, of which:	19.2	Public consumption	11.7
manufacturing	13.0	Investment	16.7
Services	56.3	Exports	8.8
		Imports	-19.7

Structure of employment

	% of total		% of labour force
Agriculture	36.7	Unemployed 2019	4.1
Industry	25.3	Av. ann. rate 2009-19	4.5
Services	38.0		

Energy

	m TOE		
Total output	47.0	Net energy imports as %	
Total consumption	80.3	of energy use	42
Consumption per head			
kg oil equivalent	408		

Inflation and finance

Consumer price			% change 2018–19
inflation 2019	6.7%	Narrow money (M1)	19.2
Av. ann. inflation 2014–19	4.4%	Broad money	12.2
Deposit rate, Dec. 2019	10.37%		

Exchange rates

	end 2019		December 2019
PRs per $	154.87	Effective rates	2010 = 100
PRs per SDR	214.15	– nominal	64.4
PRs per €	174.01	– real	99.0

Trade

Principal exports	$bn fob	Principal imports	$bn cif
Knitwear	2.7	Petroleum products	7.1
Cotton fabrics	2.1	Crude oil	5.8
Rice	1.8	Palm oil	2.0
Cotton yard & thread	1.3	Telecoms equipment	1.4
Total incl. others	**23.4**	Total incl. others	**60.1**

Main export destinations	% of total	Main origins of imports	% of total
United States	16.0	China	23.7
China	7.9	United Arab Emirates	14.4
United Kingdom	7.3	Saudi Arabia	5.3
Germany	5.5	United States	4.8

Balance of payments, reserves and debt, $bn

Visible exports fob	24.8	Change in reserves	-6.6
Visible imports fob	-57.7	Level of reserves	
Trade balance	-32.8	end Dec.	11.8
Invisibles inflows	5.9	No. months of import cover	1.9
Invisibles outflows	-16.7	Official gold holdings, m oz	2.1
Net transfers	24.1	Foreign debt	91.0
Current account balance	-19.5	– as % of GDP	28.9
– as % of GDP	-6.2	– as % of total exports	175.6
Capital balance	13.5	Debt service ratio	11.8
Overall balance	-6.3		

Health and education

Health spending, % of GDP	2.9	Education spending, % of GDP	2.9
Doctors per 1,000 pop.	1.0	Enrolment, %: primary	94
Hospital beds per 1,000 pop.	0.6	secondary	43
At least basic drinking water,		tertiary	9
% of pop.	91.5		

Society

No. of households, m	29.4	Cost of living, Dec. 2019	
Av. no. per household	7.2	New York = 100	35
Marriages per 1,000 pop.	...	Cars per 1,000 pop.	14
Divorces per 1,000 pop.	...	Telephone lines per 100 pop.	1.3
Religion, % of pop.		Mobile telephone subscribers	
Muslim	96.4	per 100 pop.	72.6
Hindu	1.9	Internet access, %	15.5
Christian	1.6	Broadband subs per 100 pop.	0.9
Jewish	<0.1	Broadband speed, Mbps	1.3
Non-religious	<0.1		
Other	<0.1		

PERU

Area, sq km	1,285,216	Capital	Lima
Arable as % of total land	2.7	Currency	Nuevo Sol (new Sol)

People

Population, m	32.0	Life expectancy: men	74.9 yrs
Pop. per sq km	24.9	women	80.2 yrs
Average annual rate of change		Adult literacy	94.4
in pop. 2015–20, %	1.6	Fertility rate (per woman)	2.3
Pop. aged 0–19, 2020, %	32.1	Urban population, %	77.9
Pop. aged 65 and over, 2020, %	8.7		per 1,000 pop.
No. of men per 100 women	98.7	Crude birth rate	18.1
Human Development Index	75.9	Crude death rate	5.9

The economy

GDP	$222bn	GDP per head	$6,941
GDP	New Soles 741bn	GDP per head in purchasing	
Av. ann. growth in real		power parity (USA=100)	23.0
GDP 2013–18	3.2%	Economic freedom index	67.9

Origins of GDP		**Components of GDP**	
	% of total		% of total
Agriculture	6.9	Private consumption	63.7
Industry, of which:	31.4	Public consumption	13.1
manufacturing	12.9	Investment	21.3
Services	61.6	Exports	25.4
		Imports	-23.5

Structure of employment

	% of total		% of labour force
Agriculture	27.4	Unemployed 2019	3.4
Industry	15.3	Av. ann. rate 2009–19	3.3
Services	57.3		

Energy

	m TOE		
Total output	27.7	Net energy imports as %	
Total consumption	28.2	of energy use	2
Consumption per head			
kg oil equivalent	877		

Inflation and finance

			% change 2018–19
Consumer price			
inflation 2019	2.1%	Narrow money (M1)	5.4
Av. ann. inflation 2014–19	2.7%	Broad money	9.8
Policy rate, Dec. 2019	2.25%		

Exchange rates

	end 2019		December 2019
			2010 = 100
New Soles per $	3.31	Effective rates	
New Soles per SDR	4.58	– nominal	...
New Soles per €	3.72	– real	...

Trade

Principal exports		**Principal imports**	
	$bn fob		*$bn cif*
Copper	14.9	Intermediate goods	20.5
Gold	8.3	Capital goods	11.6
Zinc	2.6	Consumer goods	9.6
Fishmeal	1.9		
Total incl. others	**49.1**	Total incl. others	**41.9**

Main export destinations		**Main origins of imports**	
	% of total		*% of total*
China	27.0	China	24.8
United States	16.3	United States	20.9
India	5.1	Brazil	6.0
South Korea	5.0	Mexico	4.4

Balance of payments, reserves and debt, $bn

Visible exports fob	49.1	Change in reserves	-3.5
Visible imports fob	-41.6	Level of reserves	
Trade balance	7.5	end Dec.	60.3
Invisibles inflows	9.1	No. months of import cover	11.1
Invisibles outflows	-23.7	Official gold holdings, m oz	1.1
Net transfers	3.6	Foreign debt	66.7
Current account balance	-3.6	– as % of GDP	29.6
– as % of GDP	-1.6	– as % of total exports	108.7
Capital balance	1.7	Debt service ratio	10.7
Overall balance	-3.4		

Health and education

Health spending, % of GDP	5.0	Education spending, % of GDP	3.7
Doctors per 1,000 pop.	1.3	Enrolment, %: primary	107
Hospital beds per 1,000 pop.	1.6	secondary	106
At least basic drinking water,		tertiary	71
% of pop.	91.1		

Society

No. of households, m	8.4	Cost of living, Dec. 2019	
Av. no. per household	3.8	New York = 100	66
Marriages per 1,000 pop.	2.5	Cars per 1,000 pop.	49
Divorces per 1,000 pop.	0.5	Telephone lines per 100 pop.	9.8
Religion, % of pop.		Mobile telephone subscribers	
Christian	95.5	per 100 pop.	123.8
Non-religious	3.0	Internet access, %	52.5
Other	1.5	Broadband subs per 100 pop.	7.4
Hindu	<0.1	Broadband speed, Mbps	3.3
Jewish	<0.1		
Muslim	<0.1		

PHILIPPINES

Area, sq km	300,000	Capital	Manila
Arable as % of total land	18.7	Currency	Philippine peso (P)

People

Population, m	106.7	Life expectancy: men	67.7 yrs
Pop. per sq km	355.7	women	75.9 yrs
Average annual rate of change		Adult literacy	98.2
in pop. 2015–20, %	1.4	Fertility rate (per woman)	2.6
Pop. aged 0–19, 2020, %	39.6	Urban population, %	46.9
Pop. aged 65 and over, 2020, %	5.5		per 1,000 pop.
No. of men per 100 women	100.9	Crude birth rate	20.6
Human Development Index	71.2	Crude death rate	6.2

The economy

GDP	$331bn	GDP per head	$3,103
GDP	P17,426bn	GDP per head in purchasing	
Av. ann. growth in real		power parity (USA=100)	14.3
GDP 2013–18	6.4%	Economic freedom index	64.5

Origins of GDP

Components of GDP

	% of total		% of total
Agriculture	9.3	Private consumption	73.8
Industry, of which:	30.7	Public consumption	11.9
manufacturing	19.1	Investment	26.9
Services	60.0	Exports	31.7
		Imports	-44.4

Structure of employment

	% of total		% of labour force
Agriculture	23.4	Unemployed 2019	2.3
Industry	19.4	Av. ann. rate 2009-19	2.2
Services	57.2		

Energy

	m TOE		
Total output	15.6	Net energy imports as %	
Total consumption	46.2	of energy use	66
Consumption per head			
kg oil equivalent	440		

Inflation and finance

Consumer price			% change 2018–19
inflation 2019	2.5%	Narrow money (M1)	-3.0
Av. ann. inflation 2014–19	2.5%	Broad money	7.0
Deposit rate, Dec. 2019	3.26%		

Exchange rates

	end 2019		December 2019
P per $	50.74	Effective rates	2010 = 100
P per SDR	70.17	– nominal	98.4
P per €	57.01	– real	108.2

Trade

Principal exports

	$bn fob
Electrical & electronic equip.	38.1
Machinery & transport equip.	5.0
Mineral products	4.1
Agricultural products	3.6
Total incl. others	**69.3**

Principal imports

	$bn cif
Raw materials & intermediate goods	44.6
Capital goods	35.4
Consumer goods	17.8
Mineral fuels & lubricants	13.7
Total incl. others	**119.3**

Main export destinations

	% of total
United States	15.2
Hong Kong	13.8
Japan	13.7
China	12.6

Main origins of imports

	% of total
China	17.9
South Korea	9.4
Japan	8.8
United States	6.6

Balance of payments, reserves and debt, $bn

Visible exports fob	52.0	Change in reserves	-2.2
Visible imports fob	-103.0	Level of reserves	
Trade balance	-51.0	end Dec.	79.2
Invisibles inflows	50.5	No. months of import cover	6.9
Invisibles outflows	-35.1	Official gold holdings, m oz	6.4
Net transfers	26.8	Foreign debt	78.8
Current account balance	-8.7	– as % of GDP	23.9
– as % of GDP	-2.6	– as % of total exports	57.8
Capital balance	8.7	Debt service ratio	6.4
Overall balance	-2.3		

Health and education

Health spending, % of GDP	4.4	Education spending, % of GDP	...
Doctors per 1,000 pop.	0.6	Enrolment, %: primary	108
Hospital beds per 1,000 pop.	1.0	secondary	86
At least basic drinking water,		tertiary	35
% of pop.	93.6		

Society

No. of households, m	23.9	Cost of living, Dec. 2019	
Av. no. per household	4.5	New York = 100	57
Marriages per 1,000 pop.	...	Cars per 1,000 pop.	32
Divorces per 1,000 pop.	...	Telephone lines per 100 pop.	3.9
Religion, % of pop.		Mobile telephone subscribers	
Christian	92.6	per 100 pop.	126.2
Muslim	5.5	Internet access, %	60.1
Other	1.7	Broadband subs per 100 pop.	3.7
Non-religious	0.1	Broadband speed, Mbps	5.2
Hindu	<0.1		
Jewish	<0.1		

POLAND

Area, sq km	312,680	Capital	Warsaw
Arable as % of total land	35.3	Currency	Zloty (Zl)

People

Population, m	37.9	Life expectancy: men	75.5 yrs
Pop. per sq km	121.1	women	83.0 yrs
Average annual rate of change		Adult literacy	...
in pop. 2015–20, %	-0.1	Fertility rate (per woman)	1.4
Pop. aged 0–19, 2020, %	19.8	Urban population, %	60.1
Pop. aged 65 and over, 2020, %	18.7		per 1,000 pop.
No. of men per 100 women	94.0	Crude birth rate	9.9
Human Development Index	87.2	Crude death rate	10.7

The economy

GDP	$586bn	GDP per head	$15,421
GDP	Zl2,116bn	GDP per head in purchasing	
Av. ann. growth in real		power parity (USA=100)	49.9
GDP 2013–18	4.1%	Economic freedom index	69.1

Origins of GDP		**Components of GDP**	
	% of total		% of total
Agriculture	2.4	Private consumption	58.1
Industry, of which:	32.7	Public consumption	17.8
manufacturing	19.1	Investment	20.7
Services	64.9	Exports	55.6
		Imports	-52.2

Structure of employment

	% of total		% of labour force
Agriculture	9.2	Unemployed 2019	3.8
Industry	31.9	Av. ann. rate 2009–19	3.5
Services	58.8		

Energy

	m TOE		
Total output	62.7	Net energy imports as %	
Total consumption	110.1	of energy use	43
Consumption per head			
kg oil equivalent	2,884		

Inflation and finance

			% change 2018–19
Consumer price			
inflation 2019	2.3%	Narrow money (M1)	4.4
Av. ann. inflation 2014–19	0.9%	Broad money	8.3
Deposit rate, Sep. 2019	0.14%		

Exchange rates

	end 2019		December 2019
Zl per $	3.80	Effective rates	2010 = 100
Zl per SDR	5.25	– nominal	97.9
Zl per €	4.27	– real	92.9

Trade

Principal exports		Principal imports	
	$bn fob		*$bn cif*
Machinery & transport equip.	97.6	Machinery & transport equip.	92.7
Manufactured goods	48.9	Manufactured goods	46.7
Foodstuffs & live animals	28.2	Chemicals & mineral products	36.4
Total incl. others	**261.8**	Total incl. others	**267.7**

Main export destinations		Main origins of imports	
	% of total		*% of total*
Germany	28.4	Germany	27.7
Czech Republic	6.4	China	7.9
United Kingdom	6.3	Russia	7.0
France	5.6	Netherlands	5.6
EU28	80.6	EU28	70.0

Balance of payments, reserves and debt, $bn

Visible exports fob	256.0	Change in reserves	3.7
Visible imports fob	-261.6	Level of reserves	
Trade balance	-5.6	end Dec.	116.9
Invisibles inflows	83.9	No. months of import cover	4.1
Invisibles outflows	-82.4	Official gold holdings, m oz	4.1
Net transfers	-1.7	Foreign debt	229.0
Current account balance	-5.8	– as % of GDP	39.1
– as % of GDP	-1.0	– as % of total exports	66.0
Capital balance	17.5	Debt service ratio	6.9
Overall balance	7.2		

Health and education

Health spending, % of GDP	6.5	Education spending, % of GDP	4.6
Doctors per 1,000 pop.	2.4	Enrolment, %: primary	100
Hospital beds per 1,000 pop.	6.5	secondary	110
At least basic drinking water,		tertiary	68
% of pop.	99.7		

Society

No. of households, m	14.6	Cost of living, Dec. 2019	
Av. no. per household	2.6	New York = 100	54
Marriages per 1,000 pop.	5.1	Cars per 1,000 pop.	580
Divorces per 1,000 pop.	1.7	Telephone lines per 100 pop.	17.3
Religion, % of pop.		Mobile telephone subscribers	
Christian	94.3	per 100 pop.	134.8
Non-religious	5.6	Internet access, %	77.5
Hindu	<0.1	Broadband subs per 100 pop.	16.1
Jewish	<0.1	Broadband speed, Mbps	19.7
Muslim	<0.1		
Other	<0.1		

PORTUGAL

Area, sq km	92,225	Capital	Lisbon
Arable as % of total land	10.7	Currency	Euro (€)

People

Population, m	10.3	Life expectancy: men	79.8 yrs
Pop. per sq km	111.7	women	85.3 yrs
Average annual rate of change		Adult literacy	96.1
in pop. 2015–20, %	-0.3	Fertility rate (per woman)	1.3
Pop. aged 0–19, 2020, %	18.2	Urban population, %	65.2
Pop. aged 65 and over, 2020, %	22.8		per 1,000 pop.
No. of men per 100 women	89.8	Crude birth rate	7.8
Human Development Index	85.0	Crude death rate	11.1

The economy

GDP	$241bn	GDP per head	$23,408
GDP	€204bn	GDP per head in purchasing	
Av. ann. growth in real		power parity (USA=100)	53.2
GDP 2013–18	2.1%	Economic freedom index	67.0

Origins of GDP

Components of GDP

	% of total		% of total
Agriculture	2.4	Private consumption	64.8
Industry, of which:	22.2	Public consumption	17.0
manufacturing	14.0	Investment	18.1
Services	75.5	Exports	43.5
		Imports	-43.4

Structure of employment

	% of total		% of labour force
Agriculture	5.9	Unemployed 2019	7.0
Industry	24.7	Av. ann. rate 2009-19	6.3
Services	69.4		

Energy

	m TOE		
Total output	5.2	Net energy imports as %	
Total consumption	26.9	of energy use	81
Consumption per head			
kg oil equivalent	2,606		

Inflation and finance

			% change 2018–19
Consumer price			
inflation 2019	0.3%	Narrow money (M1)	1.8
Av. ann. inflation 2014–19	0.8%	Broad money	5.2
Deposit rate, Dec. 2019	0.08%		

Exchange rates

	end 2019		December 2019
€ per $	0.89	Effective rates	2010 = 100
€ per SDR	1.23	– nominal	103.2
		– real	95.7

Trade

Principal exports		Principal imports	
	$bn fob		*$bn cif*
Machinery & transport equip.	19.5	Machinery & transport equip.	27.9
Food, drink & tobacco	7.2	Chemicals & related products	12.1
Chemicals & related products	5.9	Food, drink & tobacco	11.2
Mineral fuels & lubricants	4.7	Mineral fuels & lubricants	10.7
Total incl. others	**68.3**	Total incl. others	**89.0**

Main export destinations		Main origins of imports	
	% of total		*% of total*
Spain	25.4	Spain	31.4
France	12.7	Germany	13.8
Germany	11.5	France	7.6
United Kingdom	6.3	Italy	5.4
EU28	76.1	EU28	75.8

Balance of payments, reserves and debt, $bn

Visible exports fob	66.4	Overall balance	5.3
Visible imports fob	-85.1	Change in reserves	-1.1
Trade balance	-18.7	Level of reserves	
Invisibles inflows	50.1	end Dec.	24.9
Invisibles outflows	-35.5	No. months of import cover	2.5
Net transfers	4.9	Official gold holdings, m oz	12.3
Current account balance	0.9	Aid given	0.4
– as % of GDP	0.4	– as % of GNI	0.2
Capital balance	4.1		

Health and education

Health spending, % of GDP	9.0	Education spending, % of GDP	4.9
Doctors per 1,000 pop.	5.1	Enrolment, %: primary	106
Hospital beds per 1,000 pop.	3.4	secondary	120
At least basic drinking water,		tertiary	64
% of pop.	99.9		

Society

No. of households, m	4.1	Cost of living, Dec. 2019	
Av. no. per household	2.5	New York = 100	58
Marriages per 1,000 pop.	3.4	Cars per 1,000 pop.	451
Divorces per 1,000 pop.	2.1	Telephone lines per 100 pop.	49.5
Religion, % of pop.		Mobile telephone subscribers	
Christian	93.8	per 100 pop.	115.6
Non-religious	4.4	Internet access, %	74.7
Other	1.0	Broadband subs per 100 pop.	34.9
Muslim	0.6	Broadband speed, Mbps	21.3
Hindu	0.1		
Jewish	<0.1		

ROMANIA

Area, sq km	238,400	Capital	Bucharest
Arable as % of total land	37.3	Currency	Leu (RON)

People

Population, m	19.5	Life expectancy: men	73.1 yrs
Pop. per sq km	81.8	women	79.9 yrs
Average annual rate of change		Adult literacy	98.8
in pop. 2015–20, %	-0.7	Fertility rate (per woman)	1.6
Pop. aged 0–19, 2020, %	20.7	Urban population, %	54.0
Pop. aged 65 and over, 2020, %	19.2		per 1,000 pop.
No. of men per 100 women	94.6	Crude birth rate	9.8
Human Development Index	81.6	Crude death rate	13.4

The economy

GDP	$240bn	GDP per head	$12,301
GDP	RON944bn	GDP per head in purchasing	
Av. ann. growth in real		power parity (USA=100)	44.9
GDP 2013–18	4.6%	Economic freedom index	69.7

Origins of GDP		**Components of GDP**	
	% of total		% of total
Agriculture	4.8	Private consumption	62.4
Industry, of which:	31.3	Public consumption	16.6
manufacturing	20.7	Investment	24.2
Services	63.9	Exports	41.6
		Imports	-44.9

Structure of employment

	% of total		% of labour force
Agriculture	21.7	Unemployed 2019	4.2
Industry	30.1	Av. ann. rate 2009-19	4.0
Services	48.2		

Energy

	m TOE		
Total output	27.4	Net energy imports as %	
Total consumption	35.6	of energy use	23
Consumption per head			
kg oil equivalent	1,807		

Inflation and finance

			% change 2018–19
Consumer price			
inflation 2019	3.8%	Narrow money (M1)	7.9
Av. ann. inflation 2014–19	1.5%	Broad money	10.9
Deposit rate, Dec. 2019	1.94%		

Exchange rates

	end 2019		December 2019
Lei per $	4.26	Effective rates	2010 = 100
Lei per SDR	5.89	– nominal	93.9
Lei per €	4.79	– real	96.6

Trade

Principal exports	$bn fob	Principal imports	$bn cif
Machinery & transport equip.	38.1	Machinery & transport equip.	37.5
Basic metals & products	7.1	Chemical products	9.0
Textiles & apparel	4.8	Minerals, fuels & lubricants	7.9
Minerals, fuels & lubricants	3.4	Textiles & products	5.9
Total incl. others	80.0	Total incl. others	97.8

Main export destinations	% of total	Main origins of imports	% of total
Germany	23.0	Germany	20.5
Italy	11.4	Italy	9.4
France	7.1	Hungary	6.9
Hungary	4.9	Poland	5.5
EU28	77.1	EU28	74.7

Balance of payments, reserves and debt, $bn

Visible exports fob	73.1	Change in reserves	-2.3
Visible imports fob	-90.4	Level of reserves	
Trade balance	-17.4	end Dec.	42.1
Invisibles inflows	34.3	No. months of import cover	4.2
Invisibles outflows	-28.9	Official gold holdings, m oz	3.3
Net transfers	1.5	Foreign debt	112.1
Current account balance	-10.5	– as % of GDP	46.4
– as % of GDP	-4.4	– as % of total exports	100.5
Capital balance	8.0	Debt service ratio	19.4
Overall balance	-0.9		

Health and education

Health spending, % of GDP	5.2	Education spending, % of GDP	3.0
Doctors per 1,000 pop.	3.0	Enrolment, %: primary	85
Hospital beds per 1,000 pop.	6.3	secondary	91
At least basic drinking water,		tertiary	49
% of pop.	100.0		

Society

No. of households, m	7.5	Cost of living, Dec. 2019	
Av. no. per household	2.6	New York = 100	43
Marriages per 1,000 pop.	7.3	Cars per 1,000 pop.	278
Divorces per 1,000 pop.	1.6	Telephone lines per 100 pop.	18.8
Religion, % of pop.		Mobile telephone subscribers	
Christian	99.5	per 100 pop.	116.3
Muslim	0.3	Internet access, %	70.7
Non-religious	0.1	Broadband subs per 100 pop.	26.1
Hindu	<0.1	Broadband speed, Mbps	38.6
Jewish	<0.1		
Other	<0.1		

RUSSIA

Area, sq km	17,098,250	Capital	Moscow
Arable as % of total land	7.5	Currency	Rouble (Rb)

People

Population, m	145.7	Life expectancy: men	67.6 yrs
Pop. per sq km	8.5	women	78.2 yrs
Average annual rate of change		Adult literacy	98.8
in pop. 2015–20, %	0.1	Fertility rate (per woman)	1.8
Pop. aged 0–19, 2020, %	23.2	Urban population, %	74.4
Pop. aged 65 and over, 2020, %	15.5		per 1,000 pop.
No. of men per 100 women	86.4	Crude birth rate	12.8
Human Development Index	82.4	Crude death rate	13.1

The economy

GDP	$1,658bn	GDP per head	$11,289
GDP	Rb103trn	GDP per head in purchasing	
Av. ann. growth in real		power parity (USA=100)	43.9
GDP 2013–18	0.5%	Economic freedom index	61.0

Origins of GDP		**Components of GDP**	
	% of total		% of total
Agriculture	3.5	Private consumption	49.4
Industry, of which:	35.9	Public consumption	17.4
manufacturing	13.8	Investment	22.7
Services	60.6	Exports	30.7
		Imports	-20.8

Structure of employment

	% of total		% of labour force
Agriculture	5.8	Unemployed 2019	4.8
Industry	26.7	Av. ann. rate 2009–19	4.6
Services	67.6		

Energy

	m TOE		
Total output	1,552.8	Net energy imports as %	
Total consumption	828.0	of energy use	-88
Consumption per head			
kg oil equivalent	5,750		

Inflation and finance

Consumer price		% change 2018–19	
inflation 2019	4.5%	Narrow money (M1)	4.7
Av. ann. inflation 2014–19	6.6%	Broad money	5.1
Deposit rate, Dec. 2019	4.68%		

Exchange rates

	end 2019		December 2019
Rb per $	61.91	Effective rates	2010 = 100
Rb per SDR	85.60	– nominal	66.8
Rb per €	69.56	– real	88.9

Trade

Principal exports

	$bn fob
Fuels	264.5
Ores & metals	51.4
Machinery & equipment	29.0
Chemicals	25.2
Total incl.others	**449.6**

Principal imports

	$bn cif
Machinery & equipment	110.5
Chemicals	41.9
Food & agricultural products	29.0
Metals	16.9
Total incl.others	**236.8**

Main export destinations

	% of total
China	12.5
Netherlands	9.7
Germany	7.6
Belarus	4.8
EU28	45.6

Main origins of imports

	% of total
China	22.0
Germany	10.7
United States	5.3
Belarus	5.0
EU28	37.5

Balance of payments, reserves and debt, $bn

Visible exports fob	443.1	Change in reserves	36.1
Visible imports fob	-248.7	Level of reserves	
Trade balance	194.4	end Dec.	468.5
Invisibles inflows	117.6	No. months of import cover	12.8
Invisibles outflows	-189.3	Official gold holdings, m oz	67.9
Net transfers	-8.9	Foreign debt	453.9
Current account balance	113.7	– as % of GDP	27.4
– as % of GDP	6.9	– as % of total exports	79.8
Capital balance	-79.0	Debt service ratio	19.3
Overall balance	38.2		

Health and education

Health spending, % of GDP	5.3	Education spending, % of GDP	3.7
Doctors per 1,000 pop.	4.0	Enrolment, %: primary	103
Hospital beds per 1,000 pop.	8.2	secondary	103
At least basic drinking water,		tertiary	82
% of pop.	97.1		

Society

No. of households, m	57.2	Cost of living, Dec. 2019	
Av. no. per household	2.5	New York = 100	58
Marriages per 1,000 pop.	...	Cars per 1,000 pop.	327
Divorces per 1,000 pop.	...	Telephone lines per 100 pop.	20.7
Religion, % of pop.		Mobile telephone subscribers	
Christian	73.3	per 100 pop.	157.4
Non-religious	16.2	Internet access, %	80.9
Muslim	10.0	Broadband subs per 100 pop.	22.0
Jewish	0.2	Broadband speed, Mbps	13.5
Hindu	<0.1		
Other	<0.1		

SAUDI ARABIA

Area, sq km	2,149,690	Capital	Riyadh
Arable as % of total land	1.6	Currency	Riyal (SR)

People

Population, m	33.7	Life expectancy: men	74.5 yrs
Pop. per sq km	15.3	women	77.4 yrs
Average annual rate of change		Adult literacy	95.3
in pop. 2015–20, %	1.9	Fertility rate (per woman)	2.3
Pop. aged 0–19, 2020, %	31.1	Urban population, %	83.8
Pop. aged 65 and over, 2020, %	3.5		per 1,000 pop.
No. of men per 100 women	137.1	Crude birth rate	18.0
Human Development Index	85.7	Crude death rate	3.7

The economy

GDP	$787bn	GDP per head	$23,339
GDP	SR2,949bn	GDP per head in purchasing	
Av. ann. growth in real		power parity (USA=100)	88.1
GDP 2013–18	2.2%	Economic freedom index	62.4

Origins of GDP		**Components of GDP**	
	% of total		% of total
Agriculture	2.2	Private consumption	37.9
Industry, of which:	49.5	Public consumption	24.6
manufacturing	12.8	Investment	24.2
Services	48.2	Exports	39.9
		Imports	-26.7

Structure of employment

	% of total		% of labour force
Agriculture	2.4	Unemployed 2019	6.0
Industry	24.7	Av. ann. rate 2009–19	5.9
Services	72.9		

Energy

	m TOE		
Total output	726.5	Net energy imports as %	
Total consumption	277.3	of energy use	-162
Consumption per head			
kg oil equivalent	8,420		

Inflation and finance

Consumer price			% change 2018–19
inflation 2019	-1.2%	Narrow money (M1)	2.5
Av. ann. inflation 2014–19	0.7%	Broad money	7.1
Central bank policy rate, Dec. 2019	2.25%		

Exchange rates

	end 2019		December 2019
			2010 = 100
SR per $	3.75	Effective rates	
SR per SDR	5.19	– nominal	118.6
SR per €	4.21	– real	104.2

Trade

Principal exports

	$bn fob
Crude oil	167.7
Refined petroleum products	46.1
Total incl. others	**294.4**

Principal imports

	$bn cif
Machinery & transport equip.	52.2
Foodstuffs	21.4
Chemical & metal products	13.8
Total incl. others	**136.8**

Main export destinations

	% of total
China	13.0
Japan	11.1
India	8.9
South Korea	8.8

Main origins of imports

	% of total
China	16.6
United States	13.6
United Arab Emirates	8.7
Germany	5.4

Balance of payments, reserves and aid, $bn

Visible exports fob	294.4	Change in reserves	0.0
Visible imports fob	-125.6	Level of reserves	
Trade balance	168.7	end Dec.	509.4
Invisibles inflows	38.8	No. months of import cover	27.6
Invisibles outflows	-95.9	Official gold holdings, m oz	10.4
Net transfers	-41.1	Foreign debt	218.9
Current account balance	70.6	– as % of GDP	27.8
– as % of GDP	9.0	– as % of total exports	65.7
Capital balance	-69.5	Debt service ratio	6.9
Overall balance	0.4		

Health and education

Health spending, % of GDP	5.2	Education spending, % of GDP	...
Doctors per 1,000 pop.	2.6	Enrolment, %: primary	100
Hospital beds per 1,000 pop.	2.7	secondary	110
At least basic drinking water,		tertiary	68
% of pop.	100.0		

Society

No. of households, m	5.9	Cost of living, Dec. 2019	
Av. no. per household	5.2	New York = 100	52
Marriages per 1,000 pop.	...	Cars per 1,000 pop.	141
Divorces per 1,000 pop.	...	Telephone lines per 100 pop.	16.0
Religion, % of pop.		Mobile telephone subscribers	
Muslim	93.0	per 100 pop.	122.6
Christian	4.4	Internet access, %	93.3
Hindu	1.1	Broadband subs per 100 pop.	20.2
Other	0.9	Broadband speed, Mbps	4.1
Non-religious	0.7		
Jewish	<0.1		

SINGAPORE

Area, sq km	719	Capital	Singapore
Arable as % of total land	0.8	Currency	Singapore dollar (S$)

People

Population, m	5.8	Life expectancy: men	82.1 yrs
Pop. per sq km	8,078.0	women	86.2 yrs
Average annual rate of change		Adult literacy	97.3
in pop. 2015–20, %	0.9	Fertility rate (per woman)	1.2
Pop. aged 0–19, 2020, %	16.8	Urban population, %	100.0
Pop. aged 65 and over, 2020, %	13.4		per 1,000 pop.
No. of men per 100 women	109.8	Crude birth rate	8.8
Human Development Index	93.5	Crude death rate	5.2

The economy

GDP	$364bn	GDP per head	$64,582
GDP	S$491bn	GDP per head in purchasing	
Av. ann. growth in real		power parity (USA=100)	161.7
GDP 2013–18	3.3%	Economic freedom index	89.4

Origins of GDP		Components of GDP	
	% of total		% of total
Agriculture	0.0	Private consumption	34.9
Industry, of which:	26.7	Public consumption	10.5
manufacturing	21.9	Investment	26.6
Services	73.3	Exports	176.4
		Imports	-149.8

Structure of employment

	% of total		% of labour force
Agriculture	0.7	Unemployed 2019	4.0
Industry	15.5	Av. ann. rate 2009-19	4.1
Services	83.8		

Energy

	m TOE		
Total output	0.4	Net energy imports as %	
Total consumption	91.5	of energy use	100
Consumption per head			
kg oil equivalent	16,034		

Inflation and finance

			% change 2018–19
Consumer price			
inflation 2019	0.6%	Narrow money (M1)	2.9
Av. ann. inflation 2014–19	0.1%	Broad money	5.0
Deposit rate, Dec. 2019	0.20%		

Exchange rates

	end 2019		December 2019
			2010 = 100
S$ per $	1.35	Effective rates	
S$ per SDR	1.86	– nominal	113.4
S$ per €	1.52	– real	106.3

Trade

Principal exports		Principal imports	
	$bn fob		*$bn cif*
Machinery & transport equip.	195.4	Machinery & transport equip.	171.2
Mineral fuels	76.6	Mineral fuels	88.0
Construction materials & metals	59.0	Misc. manufactured articles	30.2
Chemicals & chemical products	47.5	Chemicals & chemical products	29.5
Total incl.others	**411.5**	Total incl.others	**370.2**

Main export destinations		Main origins of imports	
	% of total		*% of total*
China	12.3	China	13.4
Hong Kong	11.8	Malaysia	11.6
Malaysia	10.9	United States	11.4
Indonesia	8.0	Taiwan	8.5

Balance of payments, reserves and debt, $bn

Visible exports fob	460.5	Change in reserves	7.7
Visible imports fob	-356.5	Level of reserves	
Trade balance	104.0	end Dec.	292.7
Invisibles inflows	315.9	No. months of import cover	5.0
Invisibles outflows	-349.7	Official gold holdings, m oz	4.1
Net transfers	-6.2	Foreign debt	602.5
Current account balance	64.1	– as % of GDP	161.4
– as % of GDP	17.6	– as % of total exports	77.6
Capital balance	-47.7	Debt service ratio	7.8
Overall balance	14.0		

Health and education

Health spending, % of GDP	4.4	Education spending, % of GDP	...
Doctors per 1,000 pop.	2.3	Enrolment, %: primary	101
Hospital beds per 1,000 pop.	2.4	secondary	108
At least basic drinking water,		tertiary	85
% of pop.	100		

Society

No. of households, m	1.3	Cost of living, Dec. 2019	
Av. no. per household	4.5	New York = 100	102
Marriages per 1,000 pop.	6.8	Cars per 1,000 pop.	115
Divorces per 1,000 pop.	1.8	Telephone lines per 100 pop.	34.8
Religion, % of pop.		Mobile telephone subscribers	
Buddhist	33.9	per 100 pop.	148.8
Christian	18.2	Internet access, %	88.2
Non-religious	16.4	Broadband subs per 100 pop.	28.0
Muslim	14.3	Broadband speed, Mbps	60.4
Other	12.0		
Hindu	5.2		

SLOVAKIA

Area, sq km	49,030	Capital	Bratislava
Arable as % of total land	28.0	Currency	Euro (€)

People

Population, m	5.5	Life expectancy: men	74.6 yrs
Pop. per sq km	112.2	women	81.4 yrs
Average annual rate of change		Adult literacy	...
in pop. 2015–20, %	0.1	Fertility rate (per woman)	1.5
Pop. aged 0–19, 2020, %	20.4	Urban population, %	53.7
Pop. aged 65 and over, 2020, %	16.7		per 1,000 pop.
No. of men per 100 women	94.9	Crude birth rate	10.5
Human Development Index	85.7	Crude death rate	10.4

The economy

GDP	$106bn	GDP per head	$19,443
GDP	€90bn	GDP per head in purchasing	
Av. ann. growth in real		power parity (USA=100)	53.7
GDP 2013–18	3.4%	Economic freedom index	66.8

Origins of GDP		**Components of GDP**	
	% of total		% of total
Agriculture	2.6	Private consumption	55.9
Industry, of which:	33.6	Public consumption	18.6
manufacturing	21.9	Investment	23.4
Services	63.8	Exports	96.1
		Imports	-94.1

Structure of employment

	% of total		% of labour force
Agriculture	2.2	Unemployed 2019	6.5
Industry	36.3	Av. ann. rate 2009-19	5.6
Services	61.6		

Energy

	m TOE		
Total output	6.4	Net energy imports as %	
Total consumption	18.6	of energy use	66
Consumption per head			
kg oil equivalent	3,406		

Inflation and finance

Consumer price			% change 2018–19
inflation 2019	2.8%	Narrow money (M1)	1.8
Av. ann. inflation 2014–19	1.2%	Broad money	5.2
Deposit rate, Dec. 2019	0.78%		

Exchange rates

	end 2019		December 2019
			2010 = 100
€ per $	0.89	Effective rates	
€ per SDR	1.23	– nominal	103.2
		– real	99.9

Trade

Principal exports

	$bn fob
Machinery & transport equip.	56.7
Chemicals & related products	4.0
Food, drink & tobacco	2.9
Mineral fuels & lubricants	2.7
Total incl. others	**93.5**

Principal imports

	$bn cif
Machinery & transport equip.	46.0
Chemicals & related products	7.7
Mineral fuels & lubricants	6.9
Food, drink & tobacco	4.9
Total incl. others	**90.7**

Main export destinations

	% of total
Germany	22.2
Czech Republic	11.9
Poland	7.8
France	6.3
EU28	85.5

Main origins of imports

	% of total
Germany	20.4
Czech Republic	16.3
Austria	10.3
Poland	7.1
EU28	79.7

Balance of payments, reserves and debt, $bn

Visible exports fob	89.4	Overall balance	0.3
Visible imports fob	-89.6	Change in reserves	1.6
Trade balance	-0.2	Level of reserves	
Invisibles inflows	16.3	end Dec.	5.2
Invisibles outflows	-17.4	No. months of import cover	0.6
Net transfers	-1.5	Official gold holdings, m oz	1.0
Current account balance	-2.8	Aid given	0.1
– as % of GDP	-2.6	– as % of GNI	0.1
Capital balance	4.0		

Health and education

Health spending, % of GDP	6.7	Education spending, % of GDP	3.9
Doctors per 1,000 pop.	3.4	Enrolment, %: primary	99
Hospital beds per 1,000 pop.	5.8	secondary	91
At least basic drinking water,		tertiary	47
% of pop.	99.8		

Society

No. of households, m	1.8	Cost of living, Dec. 2019	
Av. no. per household	3.1	New York = 100	…
Marriages per 1,000 pop.	5.7	Cars per 1,000 pop.	395
Divorces per 1,000 pop.	1.8	Telephone lines per 100 pop.	13.3
Religion, % of pop.		Mobile telephone subscribers	
Christian	85.3	per 100 pop.	132.8
Non-religious	14.3	Internet access, %	80.7
Muslim	0.2	Broadband subs per 100 pop.	27.7
Other	0.1	Broadband speed, Mbps	25.3
Hindu	<0.1		
Jewish	<0.1		

SLOVENIA

Area, sq km	20,675	Capital	Ljubljana
Arable as % of total land	9.1	Currency	Euro (€)

People

Population, m	2.1	Life expectancy: men	79.3 yrs
Pop. per sq km	103.6	women	84.4 yrs
Average annual rate of change		Adult literacy	99.7
in pop. 2015–20, %	0.1	Fertility rate (per woman)	1.6
Pop. aged 0–19, 2020, %	19.6	Urban population, %	54.5
Pop. aged 65 and over, 2020, %	20.7		per 1,000 pop.
No. of men per 100 women	99.2	Crude birth rate	9.7
Human Development Index	90.2	Crude death rate	10.5

The economy

GDP	$54bn	GDP per head	$26,124
GDP	€46bn	GDP per head in purchasing	
Av. ann. growth in real		power parity (USA=100)	60.6
GDP 2013–18	3.4%	Economic freedom index	67.8

Origins of GDP		Components of GDP	
	% of total		% of total
Agriculture	2.4	Private consumption	52.3
Industry, of which:	32.7	Public consumption	18.3
manufacturing	23.5	Investment	21.1
Services	64.9	Exports	85.4
		Imports	-77.1

Structure of employment

	% of total		% of labour force
Agriculture	5.2	Unemployed 2019	5.1
Industry	33.5	Av. ann. rate 2009–19	4.2
Services	61.3		

Energy

	m TOE		
Total output	3.5	Net energy imports as %	
Total consumption	7.2	of energy use	51
Consumption per head			
kg oil equivalent	3,445		

Inflation and finance

		% change 2018–19	
Consumer price			
inflation 2019	1.6%	Narrow money (M1)	1.8
Av. ann. inflation 2014–19	0.8%	Broad money	5.2
Deposit rate, Dec. 2019	0.17%		

Exchange rates

	end 2019		December 2019
€ per $	0.89	Effective rates	2010 = 100
€ per SDR	1.23	– nominal	...
		– real	...

Trade

Principal exports

	$bn fob
Machinery & transport equip.	17.9
Manufactures	13.4
Chemicals	6.6
Miscellaneous manufactures	6.3
Total incl. others	**44.2**

Principal imports

	$bn cif
Machinery & transport equip.	15.3
Manufactures	11.9
Chemicals	6.0
Mineral fuels & lubricants	3.8
Total incl. others	**42.3**

Main export destinations

	% of total
Germany	18.5
Italy	11.6
Austria	7.4
Croatia	7.1
EU28	76.3

Main origins of imports

	% of total
Germany	15.5
Italy	12.9
Austria	9.0
Turkey	6.0
EU28	67.2

Balance of payments, reserves and debt, $bn

Visible exports fob	36.8	Overall balance	0.1
Visible imports fob	-35.4	Change in reserves	0.0
Trade balance	1.3	Level of reserves	
Invisibles inflows	11.3	end Dec.	0.9
Invisibles outflows	-9.1	No. months of import cover	0.3
Net transfers	-0.5	Official gold holdings, m oz	0.1
Current account balance	3.1	Aid given	0.1
– as % of GDP	5.7	– as % of GNI	0.2
Capital balance	-3.2		

Health and education

Health spending, % of GDP	8.2	Education spending, % of GDP	4.8
Doctors per 1,000 pop.	3.1	Enrolment, %: primary	100
Hospital beds per 1,000 pop.	4.6	secondary	116
At least basic drinking water,		tertiary	79
% of pop.	99.5		

Society

No. of households, m	0.9	Cost of living, Dec. 2019	
Av. no. per household	2.3	New York = 100	...
Marriages per 1,000 pop.	3.5	Cars per 1,000 pop.	531
Divorces per 1,000 pop.	1.1	Telephone lines per 100 pop.	33.4
Religion, % of pop.		Mobile telephone subscribers	
Christian	78.4	per 100 pop.	118.7
Non-religious	18.0	Internet access, %	79.8
Muslim	3.6	Broadband subs per 100 pop.	29.5
Hindu	<0.1	Broadband speed, Mbps	21.4
Jewish	<0.1		
Other	<0.1		

SOUTH AFRICA

Area, sq km	1,219,090	Capital	Pretoria
Arable as % of total land	10.3	Currency	Rand (R)

People

Population, m	57.8	Life expectancy: men	61.5 yrs
Pop. per sq km	47.6	women	68.4 yrs
Average annual rate of change		Adult literacy	87.0
in pop. 2015–20, %	1.4	Fertility rate (per woman)	2.4
Pop. aged 0–19, 2020, %	37.1	Urban population, %	66.4
Pop. aged 65 and over, 2020, %	5.5		per 1,000 pop.
No. of men per 100 women	97.1	Crude birth rate	20.7
Human Development Index	70.5	Crude death rate	9.4

The economy

GDP	$368bn	GDP per head	$6,374
GDP	R4,874bn	GDP per head in purchasing	
Av. ann. growth in real		power parity (USA=100)	21.8
GDP 2013–18	1.1%	Economic freedom index	58.8

Origins of GDP		**Components of GDP**	
	% of total		% of total
Agriculture	2.4	Private consumption	59.9
Industry, of which:	29.0	Public consumption	21.3
manufacturing	13.2	Investment	17.9
Services	68.5	Exports	29.9
		Imports	-29.6

Structure of employment

	% of total		% of labour force
Agriculture	5.1	Unemployed 2019	26.9
Industry	22.9	Av. ann. rate 2009-19	28.2
Services	72.0		

Energy

	m TOE		
Total output	149.7	Net energy imports as %	
Total consumption	143.1	of energy use	-5
Consumption per head			
kg oil equivalent	2,524		

Inflation and finance

			% change 2018–19
Consumer price			
inflation 2019	4.1%	Narrow money (M1)	4.9
Av. ann. inflation 2014–19	5.0%	Broad money	6.1
Deposit rate, Dec. 2019	6.81%		

Exchange rates

	end 2019		December 2019
R per $	14.03	Effective rates	2010 = 100
R per SDR	19.39	– nominal	61.9
R per €	15.76	– real	78.9

Trade

Principal exports		Principal imports	
	$bn fob		*$bn cif*
Mineral products	22.3	Machinery & equipment	20.1
Precious metals	16.6	Mineral products	17.6
Vehicles, aircraft & vessels	11.6	Chemicals	10.2
Iron & steel products	11.4	Vehicles, aircraft & vessels	8.1
Total incl. others	**94.1**	Total incl. others	**93.1**

Main export destinations		Main origins of imports	
	% of total		*% of total*
China	9.2	China	18.4
Germany	7.5	Germany	9.9
United States	6.7	United States	6.0
United Kingdom	5.1	Saudi Arabia	5.8

Balance of payments, reserves and debt, $bn

Visible exports fob	94.1	Change in reserves	0.9
Visible imports fob	-92.4	Level of reserves	
Trade balance	1.7	end Dec.	51.6
Invisibles inflows	23.3	No. months of import cover	4.8
Invisibles outflows	-35.6	Official gold holdings, m oz	4.0
Net transfers	-2.7	Foreign debt	179.3
Current account balance	-13.4	– as % of GDP	48.7
– as % of GDP	-3.6	– as % of total exports	164.5
Capital balance	11.9	Debt service ratio	21.4
Overall balance	0.8		

Health and education

Health spending, % of GDP	8.1	Education spending, % of GDP	6.2
Doctors per 1,000 pop.	0.9	Enrolment, %: primary	101
Hospital beds per 1,000 pop.	2.8	secondary	105
At least basic drinking water,		tertiary	22
% of pop.	92.7		

Society

No. of households, m	18.0	Cost of living, Dec. 2019	
Av. no. per household	3.2	New York = 100	51
Marriages per 1,000 pop.	...	Cars per 1,000 pop.	115
Divorces per 1,000 pop.	...	Telephone lines per 100 pop.	5.8
Religion, % of pop.		Mobile telephone subscribers	
Christian	81.2	per 100 pop.	159.9
Non-religious	14.9	Internet access, %	56.2
Muslim	1.7	Broadband subs per 100 pop.	1.9
Hindu	1.1	Broadband speed, Mbps	6.4
Other	0.9		
Jewish	0.1		

SOUTH KOREA

Area, sq km	100,339	Capital	Seoul
Arable as % of total land	14.6	Currency	Won (W)

People

Population, m	51.2	Life expectancy: men	80.5 yrs
Pop. per sq km	510.6	women	86.4 yrs
Average annual rate of change		Adult literacy	...
in pop. 2015–20, %	0.2	Fertility rate (per woman)	1.1
Pop. aged 0–19, 2020, %	17.4	Urban population, %	81.5
Pop. aged 65 and over, 2020, %	15.8		per 1,000 pop.
No. of men per 100 women	100.2	Crude birth rate	7.4
Human Development Index	90.6	Crude death rate	6.9

The economy

GDP	$1,619bn	GDP per head	$31,363
GDP	W1,893,497bn	GDP per head in purchasing	
Av. ann. growth in real		power parity (USA=100)	63.9
GDP 2013–18	3.0%	Economic freedom index	74.0

Origins of GDP		**Components of GDP**	
	% of total		% of total
Agriculture	2.0	Private consumption	48.0
Industry, of which:	37.3	Public consumption	16.1
manufacturing	29.2	Investment	31.3
Services	60.7	Exports	41.6
		Imports	-37.0

Structure of employment

	% of total		% of labour force
Agriculture	4.9	Unemployed 2019	3.8
Industry	25.1	Av. ann. rate 2009-19	4.1
Services	70.0		

Energy

	m TOE		
Total output	40.9	Net energy imports as %	
Total consumption	311.7	of energy use	87
Consumption per head			
kg oil equivalent	6,114		

Inflation and finance

		% change 2018–19	
Consumer price			
inflation 2019	0.4%	Narrow money (M1)	11.3
Av. ann. inflation 2014–19	1.1%	Broad money	7.9
Deposit rate, Dec. 2019	1.68%		

Exchange rates

	end 2019		December 2019
W per $	1,157.8	Effective rates	2010 = 100
W per SDR	1,601.0	– nominal	...
W per €	1,300.9	– real	...

Trade

Principal exports		Principal imports	
	$bn fob		*$bn cif*
Machinery & transport equip.	345.4	Machinery & transport equip.	166.2
Chemicals & related products	80.7	Mineral fuels & lubricants	147.0
Manufactured goods	73.9	Chemicals & related products	55.2
Mineral fuels & lubricants	48.0	Manufactured goods	52.0
Total incl. others	**605.0**	Total incl. others	**535.5**

Main export destinations		Main origins of imports	
	% of total		*% of total*
China	26.8	China	19.9
United States	12.1	United States	11.0
Vietnam	8.0	Japan	10.2
Hong Kong	7.6	Saudi Arabia	4.9

Balance of payments, reserves and debt, $bn

Visible exports fob	626.3	Change in reserves	14.3
Visible imports fob	-516.2	Level of reserves	
Trade balance	110.1	end Dec.	403.1
Invisibles inflows	138.5	No. months of import cover	7.1
Invisibles outflows	-163.0	Official gold holdings, m oz	3.4
Net transfers	-8.2	Foreign debt	404.8
Current account balance	77.5	– as % of GDP	23.5
– as % of GDP	4.8	– as % of total exports	52.9
Capital balance	-59.1	Debt service ratio	5.4
Overall balance	17.5		

Health and education

Health spending, % of GDP	7.6	Education spending, % of GDP	4.6
Doctors per 1,000 pop.	2.4	Enrolment, %: primary	98
Hospital beds per 1,000 pop.	11.5	secondary	100
At least basic drinking water,		tertiary	94
% of pop.	99.8		

Society

No. of households, m	20.4	Cost of living, Dec. 2019	
Av. no. per household	2.5	New York = 100	90
Marriages per 1,000 pop.	5.0	Cars per 1,000 pop.	339
Divorces per 1,000 pop.	2.1	Telephone lines per 100 pop.	50.6
Religion, % of pop.		Mobile telephone subscribers	
Non-religious	46.4	per 100 pop.	129.7
Christian	29.4	Internet access, %	96.0
Buddhist	22.9	Broadband subs per 100 pop.	41.6
Other	1.0	Broadband speed, Mbps	20.6
Muslim	0.2		
Jewish	<0.1		

SPAIN

Area, sq km	505,992	Capital	Madrid
Arable as % of total land	24.7	Currency	Euro (€)

People

Population, m	46.7	Life expectancy: men	81.3 yrs
Pop. per sq km	92.3	women	86.7 yrs
Average annual rate of change		Adult literacy	98.4
in pop. 2015–20, %	0.0	Fertility rate (per woman)	1.3
Pop. aged 0–19, 2020, %	19.2	Urban population, %	80.3
Pop. aged 65 and over, 2020, %	20.0		per 1,000 pop.
No. of men per 100 women	96.6	Crude birth rate	8.5
Human Development Index	89.3	Crude death rate	9.5

The economy

GDP	$1,419bn	GDP per head	$30,371
GDP	€1,208bn	GDP per head in purchasing	
Av. ann. growth in real		power parity (USA=100)	63.2
GDP 2013–18	2.7%	Economic freedom index	66.9

Origins of GDP		**Components of GDP**	
	% of total		% of total
Agriculture	3.1	Private consumption	58.3
Industry, of which:	22.1	Public consumption	18.6
manufacturing	12.4	Investment	20.4
Services	74.8	Exports	35.1
		Imports	-32.4

Structure of employment

	% of total		% of labour force
Agriculture	4.1	Unemployed 2019	15.3
Industry	20.3	Av. ann. rate 2009–19	14
Services	75.6		

Energy

	m TOE		
Total output	35.4	Net energy imports as %	
Total consumption	114.6	of energy use	76
Consumption per head			
kg oil equivalent	3,120		

Inflation and finance

			% change 2018–19
Consumer price			
inflation 2019	0.7%	Narrow money (M1)	1.8
Av. ann. inflation 2014–19	0.7%	Broad money	5.2
Deposit rate, Dec. 2019	0.03%		

Exchange rates

	end 2019		December 2019
			2010 = 100
€ per $	0.89	Effective rates	
€ per SDR	1.23	– nominal	103.9
		– real	94.8

Trade

Principal exports

	$bn fob
Machinery & transport equip.	111.9
Food, drink & tobacco	51.8
Chemicals & related products	46.1
Mineral fuels & lubricants	26.9
Total incl. others	**337.1**

Principal imports

	$bn cif
Machinery & transport equip.	120.9
Mineral fuels & lubricants	56.4
Chemicals & related products	56.0
Food, drink & tobacco	38.4
Total incl. others	**377.6**

Main export destinations

	% of total
France	15.2
Germany	11.0
Italy	8.0
Portugal	7.5
EU28	66.4

Main origins of imports

	% of total
Germany	14.1
France	11.4
China	7.1
Italy	7.0
EU28	58.8

Balance of payments, reserves and aid, $bn

Visible exports fob	343.0	Overall balance	2.6
Visible imports fob	-377.5	Change in reserves	1.3
Trade balance	-34.5	Level of reserves	
Invisibles inflows	225.0	end Dec.	70.6
Invisibles outflows	-149.0	No. months of import cover	1.6
Net transfers	-14.2	Official gold holdings, m oz	9.1
Current account balance	27.3	Aid given	2.5
– as % of GDP	1.9	– as % of GNI	0.2
Capital balance	-27.1		

Health and education

Health spending, % of GDP	8.9	Education spending, % of GDP	4.2
Doctors per 1,000 pop.	3.9	Enrolment, %: primary	103
Hospital beds per 1,000 pop.	3.0	secondary	126
At least basic drinking water,		tertiary	89
% of pop.	99.9		

Society

No. of households, m	18.6	Cost of living, Dec. 2019	
Av. no. per household	2.5	New York = 100	74
Marriages per 1,000 pop.	3.5	Cars per 1,000 pop.	483
Divorces per 1,000 pop.	2.0	Telephone lines per 100 pop.	41.7
Religion, % of pop.		Mobile telephone subscribers	
Christian	78.6	per 100 pop.	116.0
Non-religious	19.0	Internet access, %	86.1
Muslim	2.1	Broadband subs per 100 pop.	32.5
Jewish	0.1	Broadband speed, Mbps	27.2
Other	0.1		
Hindu	<0.1		

SWEDEN

Area, sq km	447,430	Capital	Stockholm
Arable as % of total land	6.3	Currency	Swedish krona (Skr)

People

Population, m	10.0	Life expectancy: men	81.7 yrs
Pop. per sq km	22.2	women	85.0 yrs
Average annual rate of change		Adult literacy	...
in pop. 2015–20, %	0.7	Fertility rate (per woman)	1.9
Pop. aged 0–19, 2020, %	23.0	Urban population, %	87.4
Pop. aged 65 and over, 2020, %	20.3		per 1,000 pop.
No. of men per 100 women	100.4	Crude birth rate	11.9
Human Development Index	93.7	Crude death rate	9.1

The economy

GDP	$556bn	GDP per head	$54,608
GDP	Skr4,834bn	GDP per head in purchasing	
Av. ann. growth in real		power parity (USA=100)	84.7
GDP 2013–18	2.8%	Economic freedom index	74.9

Origins of GDP		**Components of GDP**	
	% of total		% of total
Agriculture	1.6	Private consumption	44.7
Industry, of which:	24.9	Public consumption	26.0
manufacturing	15.0	Investment	26.8
Services	73.5	Exports	45.8
		Imports	-43.3

Structure of employment

	% of total		% of labour force
Agriculture	1.6	Unemployed 2019	6.3
Industry	17.9	Av. ann. rate 2009–19	6.5
Services	80.4		

Energy

	m TOE		
Total output	38.4	Net energy imports as %	
Total consumption	56.2	of energy use	32
Consumption per head			
kg oil equivalent	5,671		

Inflation and finance

			% change 2018–19
Consumer price			
inflation 2019	1.7%	Narrow money (M1)	9.0
Av. ann. inflation 2014–19	1.5%	Broad money	7.1
Repo rate, Dec. 2019	0.00%		

Exchange rates

	end 2019		December 2019
			2010 = 100
Skr per $	9.30	Effective rates	
Skr per SDR	12.86	– nominal	92.9
Skr per €	10.45	– real	86.1

Trade

Principal exports

	$bn fob
Machinery & transport equip.	66.4
Chemicals & related products	19.9
Mineral fuels & lubricants	13.2
Raw materials	11.3
Total incl. others	**165.9**

Principal imports

	$bn cif
Machinery & transport equip.	65.2
Food, drink & tobacco	20.2
Chemicals & related products	18.4
Mineral fuels & lubricants	17.3
Total incl. others	**170.2**

Main export destinations

	% of total
Germany	10.9
Norway	10.5
Finland	7.0
Denmark	6.9
EU28	59.5

Main origins of imports

	% of total
Germany	17.9
Netherlands	9.4
Norway	8.2
Denmark	7.0
EU28	70.0

Balance of payments, reserves and aid, $bn

Visible exports fob	178.5	Overall balance	0.1
Visible imports fob	-169.9	Change in reserves	-1.6
Trade balance	8.6	Level of reserves	
Invisibles inflows	129.5	end Dec.	60.6
Invisibles outflows	-119.2	No. months of import cover	2.5
Net transfers	-9.2	Official gold holdings, m oz	4.0
Current account balance	9.7	Aid given	5.8
– as % of GDP	1.7	– as % of GNI	1.0
Capital balance	-7.0		

Health and education

Health spending, % of GDP	11.0	Education spending, % of GDP	7.7
Doctors per 1,000 pop.	4.0	Enrolment, %: primary	127
Hospital beds per 1,000 pop.	2.6	secondary	153
At least basic drinking water,		tertiary	67
% of pop.	100		

Society

No. of households, m	4.6	Cost of living, Dec. 2019	
Av. no. per household	2.2	New York = 100	68
Marriages per 1,000 pop.	5.0	Cars per 1,000 pop.	476
Divorces per 1,000 pop.	2.5	Telephone lines per 100 pop.	24
Religion, % of pop.		Mobile telephone subscribers	
Christian	67.2	per 100 pop.	126.8
Non-religious	27.0	Internet access, %	92.1
Muslim	4.6	Broadband subs per 100 pop.	39.9
Other	0.8	Broadband speed, Mbps	46.0
Hindu	0.2		
Jewish	0.1		

SWITZERLAND

Area, sq km	41,290	Capital	Bern
Arable as % of total land	10.1	Currency	Swiss franc (SFr)

People

Population, m	8.5	Life expectancy: men	82.4 yrs
Pop. per sq km	205.9	women	86.0 yrs
Average annual rate of change		Adult literacy	...
in pop. 2015–20, %	0.8	Fertility rate (per woman)	1.5
Pop. aged 0–19, 2020, %	19.9	Urban population, %	73.8
Pop. aged 65 and over, 2020, %	19.1		per 1,000 pop.
No. of men per 100 women	98.5	Crude birth rate	10.3
Human Development Index	94.6	Crude death rate	8.2

The economy

GDP	$705bn	GDP per head	$82,797
GDP	SFr690bn	GDP per head in purchasing	
Av. ann. growth in real		power parity (USA=100)	108.4
GDP 2013–18	2.0%	Economic freedom index	82.0

Origins of GDP		**Components of GDP**	
	% of total		% of total
Agriculture	0.7	Private consumption	53.3
Industry, of which:	26.6	Public consumption	11.8
manufacturing	19.4	Investment	22.7
Services	72.7	Exports	66.1
		Imports	-53.9

Structure of employment

	% of total		% of labour force
Agriculture	2.9	Unemployed 2019	4.7
Industry	20.2	Av. ann. rate 2009-19	4.6
Services	76.8		

Energy

	m TOE		
Total output	14.1	Net energy imports as %	
Total consumption	29.6	of energy use	53
Consumption per head			
kg oil equivalent	3,493		

Inflation and finance

Consumer price			% change 2018–19
inflation 2019	0.4%	Narrow money (M1)	7.2
Av. ann. inflation 2014–19	0.0%	Broad money	0.1
Deposit rate, Dec. 2019	-0.39%		

Exchange rates

	end 2019		December 2019
			2010 = 100
SFr per $	0.97	Effective rates	
SFr per SDR	1.34	– nominal	124.9
SFr per €	1.09	– real	104.7

Trade

Principal exports		Principal imports	
	$bn fob		*$bn cif*
Chemicals	100.7	Chemicals	47.8
Precision instruments, watches		Machinery, equipment &	
& jewellery	47.8	electronics	31.2
Machinery, equipment &		Precision instruments, watches	
electronics	32.7	& jewellery	23.6
Metals & metal manufactures	13.9	Motor vehicles	19.4
Total incl. others	**238.4**	Total incl. others	**206.4**

Main export destinations		Main origins of imports	
	% of total		*% of total*
Germany	15.2	Germany	21.0
United States	13.2	United Kingdom	9.5
China	9.8	Italy	7.7
France	6.3	United States	7.6
EU28	44.4	EU28	63.0

Balance of payments, reserves and aid, $bn

Visible exports fob	335.5	Overall balance	13.8
Visible imports fob	-275.2	Change in reserves	-23.9
Trade balance	60.4	Level of reserves	
Invisibles inflows	303.5	end Dec.	786.9
Invisibles outflows	-296.5	No. months of import cover	16.5
Net transfers	-9.6	Official gold holdings, m oz	33.4
Current account balance	57.9	Aid given	3.1
– as % of GDP	8.2	– as % of GNI	0.5
Capital balance	-53.5		

Health and education

Health spending, % of GDP	12.3	Education spending, % of GDP	5.1
Doctors per 1,000 pop.	4.3	Enrolment, %: primary	105
Hospital beds per 1,000 pop.	4.7	secondary	102
At least basic drinking water,		tertiary	60
% of pop.	100		

Society

No. of households, m	3.8	Cost of living, Dec. 2019	
Av. no. per household	2.2	New York = 100	95
Marriages per 1,000 pop.	4.8	Cars per 1,000 pop.	540
Divorces per 1,000 pop.	1.9	Telephone lines per 100 pop.	38.7
Religion, % of pop.		Mobile telephone subscribers	
Christian	81.3	per 100 pop.	126.8
Non-religious	11.9	Internet access, %	89.7
Muslim	5.5	Broadband subs per 100 pop.	46.4
Other	0.6	Broadband speed, Mbps	29.9
Hindu	0.4		
Jewish	0.3		

TAIWAN

Area, sq km	36,179	Capital	Taipei
Arable as % of total land	...	Currency	Taiwan dollar (T$)

People

Population, m	23.7	Life expectancy: men	78.5 yrs
Pop. per sq km	655.1	women	83.6 yrs
Average annual rate of change		Adult literacy	...
in pop. 2015–20, %	0.2	Fertility rate (per woman)	1.2
Pop. aged 0–19, 2020, %	17.8	Urban population, %	77.3
Pop. aged 65 and over, 2020, %	15.8		per 1,000 pop.
No. of men per 100 women	98.8	Crude birth rate	8.4
Human Development Index	...	Crude death rate	8.1

The economy

GDP	$590bn	GDP per head	$25,008
GDP	T$17,793bn	GDP per head in purchasing	
Av. ann. growth in real		power parity (USA=100)	84.5
GDP 2013–18	2.9%	Economic freedom index	77.1

Origins of GDP		**Components of GDP**	
	% of total		% of total
Agriculture	1.6	Private consumption	52.6
Industry, of which:	35.2	Public consumption	14.3
manufacturing	30.8	Investment	21.0
Services	63.2	Exports	65.5
		Imports	-55.2

Structure of employment

	% of total		% of labour force
Agriculture	4.8	Unemployed 2019	3.7
Industry	35.6	Av. ann. rate 2009–19	4.3
Services	59.7		

Energy

	m TOE		
Total output	8.5	Net energy imports as %	
Total consumption	117.8	of energy use	93
Consumption per head			
kg oil equivalent	4,988		

Inflation and finance

		% change 2018–19	
Consumer price			
inflation 2019	0.5%	Narrow money (M1)	3.5
Av. ann. inflation 2014–19	0.7%	Broad money	4.5
Discount rate, Dec. 2019	1.38%		

Exchange rates

	end 2019		December 2019
T$ per $	30.11	Effective rates	2010 = 100
T$ per SDR	41.63	– nominal	...
T$ per €	33.83	– real	...

Trade

Principal exports

	$bn fob
Machinery & electrical equip.	183.1
Basic metals & articles	31.6
Plastic & rubber articles	25.3
Chemicals	22.2
Total incl. others	**307.9**

Principal imports

	$bn cif
Machinery & electrical equip.	108.8
Minerals	54.9
Chemicals & related products	30.4
Basic metals & articles	22.5
Total incl. others	**284.0**

Main export destinations

	% of total
China	28.8
Hong Kong	12.4
United States	11.8
Japan	6.9

Main origins of imports

	% of total
China	18.8
Japan	15.4
United States	12.1
South Korea	6.8

Balance of payments, reserves and debt, $bn

Visible exports fob	352.2	Change in reserves	10.1
Visible imports fob	-284.8	Level of reserves	
Trade balance	67.4	end Dec.	479.2
Invisibles inflows	90.0	No. months of import cover	15.7
Invisibles outflows	-82.2	Official gold holdings, m oz	13.6
Net transfers	-3.4	Foreign debt	191.2
Current account balance	71.9	– as % of GDP	31.4
– as % of GDP	12.2	– as % of total exports	43.9
Capital balance	-55.2	Debt service ratio	2.8
Overall balance	12.5		

Health and education

Health spending, % of GDP	...	Education spending, % of GDP	...
Doctors per 1,000 pop.	...	Enrolment, %: primary	...
Hospital beds per 1,000 pop.	...	secondary	...
At least basic drinking water,		tertiary	...
% of pop.	...		

Society

No. of households, m	8.6	Cost of living, Dec. 2019	
Av. no. per household	2.8	New York = 100	68
Marriages per 1,000 pop.	...	Cars per 1,000 pop.	284
Divorces per 1,000 pop.	...	Telephone lines per 100 pop.	55.5
Religion, % of pop.		Mobile telephone subscribers	
Other	60.5	per 100 pop.	123.7
Buddhist	21.3	Internet access, %	86.2
Non-religious	12.7	Broadband subs per 100 pop.	24.1
Christian	5.5	Broadband speed, Mbps	28.1
Hindu	<0.1		
Jewish	<0.1		

THAILAND

Area, sq km	513,120	Capital	Bangkok
Arable as % of total land	26.5	Currency	Baht (Bt)

People

Population, m	69.4	Life expectancy: men	74.2 yrs
Pop. per sq km	135.3	women	81.3 yrs
Average annual rate of change		Adult literacy	93.8
in pop. 2015–20, %	0.3	Fertility rate (per woman)	1.5
Pop. aged 0–19, 2020, %	22.8	Urban population, %	49.9
Pop. aged 65 and over, 2020, %	13.0		per 1,000 pop.
No. of men per 100 women	94.8	Crude birth rate	10.5
Human Development Index	76.5	Crude death rate	8.3

The economy

GDP	$505bn	GDP per head	$7,274
GDP	Bt16,318bn	GDP per head in purchasing	
Av. ann. growth in real		power parity (USA=100)	30.3
GDP 2013–18	3.1%	Economic freedom index	69.4

Origins of GDP		**Components of GDP**	
	% of total		% of total
Agriculture	8.1	Private consumption	48.7
Industry, of which:	32.3	Public consumption	16.2
manufacturing	26.7	Investment	25.0
Services	59.6	Exports	66.8
		Imports	-56.5

Structure of employment

	% of total		% of labour force
Agriculture	31.6	Unemployed 2019	0.8
Industry	22.6	Av. ann. rate 2009-19	0.8
Services	45.8		

Energy

	m TOE		
Total output	66.9	Net energy imports as %	
Total consumption	139.2	of energy use	52
Consumption per head			
kg oil equivalent	2,016		

Inflation and finance

			% change 2018–19
Consumer price			
inflation 2019	0.7%	Narrow money (M1)	3.5
Av. ann. inflation 2014–19	0.3%	Broad money	3.6
Deposit rate, Dec. 2019	1.40%		

Exchange rates

	end 2019		December 2019
Bt per $	30.15	Effective rates	2010 = 100
Bt per SDR	41.70	– nominal	...
Bt per €	33.88	– real	...

Trade

Principal exports	$bn fob	Principal imports	$bn cif
Machinery, equip. & supplies	110.4	Machinery, equip. & supplies	86.9
Manufactured goods	31.6	Manufactured goods	43.2
Food	31.3	Fuel & lubricants	41.3
Chemicals	26.6	Chemicals	27.1
Total incl. others	**250.9**	Total incl. others	**249.5**

Main export destinations	% of total	Main origins of imports	% of total
China	12.0	China	20.1
United States	11.1	Japan	14.2
Japan	9.9	United States	6.0
Vietnam	5.1	Malaysia	5.3

Balance of payments, reserves and debt, $bn

Visible exports fob	251.1	Change in reserves	3.1
Visible imports fob	-228.7	Level of reserves	
Trade balance	22.4	end Dec.	205.6
Invisibles inflows	86.1	No. months of import cover	7.8
Invisibles outflows	-88.1	Official gold holdings, m oz	5.0
Net transfers	8.0	Foreign debt	169.2
Current account balance	28.5	– as % of GDP	33.4
– as % of GDP	5.6	– as % of total exports	48.5
Capital balance	-15.6	Debt service ratio	5.3
Overall balance	7.3		

Health and education

Health spending, % of GDP	3.7	Education spending, % of GDP	...
Doctors per 1,000 pop.	0.8	Enrolment, %: primary	100
Hospital beds per 1,000 pop.	2.1	secondary	118
At least basic drinking water,		tertiary	49
% of pop.	99.9		

Society

No. of households, m	23.8	Cost of living, Dec. 2019	
Av. no. per household	5.2	New York = 100	79
Marriages per 1,000 pop.	...	Cars per 1,000 pop.	129
Divorces per 1,000 pop.	1.6	Telephone lines per 100 pop.	4.2
Religion, % of pop.		Mobile telephone subscribers	
Buddhist	93.2	per 100 pop.	180.2
Muslim	5.5	Internet access, %	56.8
Christian	0.9	Broadband subs per 100 pop.	13.2
Non-religious	0.3	Broadband speed, Mbps	17.1
Hindu	0.1		
Jewish	<0.1		

TURKEY

Area, sq km	785,350	Capital	Ankara
Arable as % of total land	26.7	Currency	Turkish Lira (YTL)

People

Population, m	82.3	Life expectancy: men	75.6 yrs
Pop. per sq km	105.0	women	81.2 yrs
Average annual rate of change		Adult literacy	96.2
in pop. 2015–20, %	1.4	Fertility rate (per woman)	2.1
Pop. aged 0–19, 2020, %	32.0	Urban population, %	75.1
Pop. aged 65 and over, 2020, %	9.0		per 1,000 pop.
No. of men per 100 women	97.5	Crude birth rate	16.2
Human Development Index	80.6	Crude death rate	5.6

The economy

GDP	$771bn	GDP per head	$9,370
GDP	YTL3,724bn	GDP per head in purchasing	
Av. ann. growth in real		power parity (USA=100)	44.7
GDP 2013–18	4.9%	Economic freedom index	64.4

Origins of GDP

Components of GDP

	% of total		% of total
Agriculture	6.5	Private consumption	56.7
Industry, of which:	32.9	Public consumption	14.8
manufacturing	21.3	Investment	29.6
Services	60.6	Exports	29.5
		Imports	-30.6

Structure of employment

	% of total		% of labour force
Agriculture	18.4	Unemployed 2019	10.9
Industry	26.3	Av. ann. rate 2009-19	13.5
Services	55.3		

Energy

	m TOE		
Total output	39.6	Net energy imports as %	
Total consumption	162.0	of energy use	76
Consumption per head			
kg oil equivalent	2,007		

Inflation and finance

		% change 2018–19	
Consumer price			
inflation 2019	15.2%	Narrow money (M1)	13.4
Av. ann. inflation 2014–19	11.6%	Broad money	27.5
Deposit rate, Dec. 2019	21.00%		

Exchange rates

	end 2019		December 2019
			2010 = 100
YTL per $	5.95	Effective rates	
YTL per SDR	8.22	– nominal	...
YTL per €	6.69	– real	...

Trade

Principal exports		Principal imports	
	$bn fob		*$bn cif*
Transport equipment	30.5	Mechanical equipment	43.0
Agricultural products	27.4	Transport equipment	33.4
Textiles & clothing	17.3	Chemicals	19.0
Iron & steel	13.2	Fuels	17.5
Total incl. others	**167.9**	Total incl. others	**223.0**

Main export destinations		Main origins of imports	
	% of total		*% of total*
Germany	9.6	Russia	9.9
United Kingdom	6.6	China	9.3
Italy	5.7	Germany	9.1
Iraq	5.0	United States	5.5
EU28	50.0	EU28	36.2

Balance of payments, reserves and debt, $bn

Visible exports fob	178.9	Change in reserves	-14.6
Visible imports fob	-219.7	Level of reserves	
Trade balance	-40.8	end Dec.	92.9
Invisibles inflows	65.2	No. months of import cover	4.2
Invisibles outflows	-46.0	Official gold holdings, m oz	15.7
Net transfers	0.9	Foreign debt	445.1
Current account balance	-20.7	– as % of GDP	57.9
– as % of GDP	-2.7	– as % of total exports	189.7
Capital balance	0.6	Debt service ratio	35.9
Overall balance	-10.4		

Health and education

Health spending, % of GDP	4.2	Education spending, % of GDP	...
Doctors per 1,000 pop.	1.8	Enrolment, %: primary	93
Hospital beds per 1,000 pop.	2.7	secondary	106
At least basic drinking water,		tertiary	...
% of pop.	98.9		

Society

No. of households, m	23.6	Cost of living, Dec. 2019	
Av. no. per household	3.5	New York = 100	54
Marriages per 1,000 pop.	7.1	Cars per 1,000 pop.	138
Divorces per 1,000 pop.	1.6	Telephone lines per 100 pop.	14.1
Religion, % of pop.		Mobile telephone subscribers	
Muslim	98.0	per 100 pop.	97.3
Non-religious	1.2	Internet access, %	71.0
Christian	0.4	Broadband subs per 100 pop.	16.3
Other	0.3	Broadband speed, Mbps	4.9
Hindu	<0.1		
Jewish	<0.1		

UKRAINE

Area, sq km	603,500	Capital	Kiev
Arable as % of total land	56.8	Currency	Hryvnya (UAH)

People

Population, m	44.2	Life expectancy: men	67.6 yrs
Pop. per sq km	73.2	women	77.3 yrs
Average annual rate of change		Adult literacy	...
in pop. 2015–20, %	-0.5	Fertility rate (per woman)	1.4
Pop. aged 0–19, 2020, %	20.5	Urban population, %	69.4
Pop. aged 65 and over, 2020, %	16.9		per 1,000 pop.
No. of men per 100 women	86.3	Crude birth rate	9.6
Human Development Index	75.0	Crude death rate	15.2

The economy

GDP	$131bn	GDP per head	$3,095
GDP	UAH3,559bn	GDP per head in purchasing	
Av. ann. growth in real		power parity (USA=100)	14.7
GDP 2013–18	-1.8%	Economic freedom index	54.9

Origins of GDP		**Components of GDP**	
	% of total		% of total
Agriculture	10.1	Private consumption	69.1
Industry, of which:	23.9	Public consumption	20.8
manufacturing	11.6	Investment	18.8
Services	66.5	Exports	45.2
		Imports	-53.8

Structure of employment

	% of total		% of labour force
Agriculture	14.5	Unemployed 2019	8.8
Industry	24.6	Av. ann. rate 2009-19	8.9
Services	61.0		

Energy

	m TOE		
Total output	58.4	Net energy imports as %	
Total consumption	91.6	of energy use	36
Consumption per head			
kg oil equivalent	2,071		

Inflation and finance

			% change 2018–19
Consumer price			
inflation 2019	7.9%	Narrow money (M1)	9.6
Av. ann. inflation 2014–19	18.3%	Broad money	12.6
Deposit rate, Dec. 2019	11.79%		

Exchange rates

	end 2019		December 2019
UAH per $	23.69	Effective rates	2010 = 100
UAH per SDR	32.75	– nominal	50.8
UAH per €	26.62	– real	99.6

Trade

Principal exports		Principal imports	
	$bn fob		$bn cif
Food & beverages	18.6	Machinery & equipment	16.5
Non-precious metals	11.6	Fuels	14.2
Machinery & equipment	5.3	Chemicals	7.0
Fuels	4.3	Food & beverages	5.1
Total incl. others	**47.3**	Total incl. others	**57.2**

Main export destinations		Main origins of imports	
	% of total		% of total
Russia	7.7	Russia	14.2
Poland	6.9	China	13.3
Italy	5.6	Germany	10.5
Turkey	5.0	Belarus	6.6
EU28	42.7	EU28	40.6

Balance of payments, reserves and debt, $bn

Visible exports fob	43.3	Change in reserves	2.0
Visible imports fob	-56.1	Level of reserves	
Trade balance	-12.7	end Dec.	20.8
Invisibles inflows	27.7	No. months of import cover	3.2
Invisibles outflows	-23.0	Official gold holdings, m oz	0.8
Net transfers	3.7	Foreign debt	114.5
Current account balance	-4.4	– as % of GDP	87.5
– as % of GDP	-3.3	– as % of total exports	133.6
Capital balance	5.5	Debt service ratio	19.7
Overall balance	2.9		

Health and education

Health spending, % of GDP	7.0	Education spending, % of GDP	5.4
Doctors per 1,000 pop.	...	Enrolment, %: primary	...
Hospital beds per 1,000 pop.	8.8	secondary	...
At least basic drinking water,		tertiary	...
% of pop.	93.8		

Society

No. of households, m	16.7	Cost of living, Dec. 2019	
Av. no. per household	2.6	New York = 100	58
Marriages per 1,000 pop.	5.9	Cars per 1,000 pop.	175
Divorces per 1,000 pop.	3.0	Telephone lines per 100 pop.	14.4
Religion, % of pop.		Mobile telephone subscribers	
Christian	83.8	per 100 pop.	127.8
Non-religious	14.7	Internet access, %	62.6
Muslim	1.2	Broadband subs per 100 pop.	12.8
Jewish	0.1	Broadband speed, Mbps	11.3
Other	0.1		
Hindu	<0.1		

UNITED ARAB EMIRATES

Area, sq km	83,600	Capital	Abu Dhabi
Arable as % of total land	0.6	Currency	Dirham (AED)

People

Population, m	9.6	Life expectancy: men	77.8 yrs
Pop. per sq km	114.8	women	79.8 yrs
Average annual rate of change		Adult literacy	93.2
in pop. 2015–20, %	1.3	Fertility rate (per woman)	1.4
Pop. aged 0–19, 2020, %	18.8	Urban population, %	86.5
Pop. aged 65 and over, 2020, %	1.3		per 1,000 pop.
No. of men per 100 women	223.8	Crude birth rate	10.4
Human Development Index	86.6	Crude death rate	1.8

The economy

GDP	$414bn	GDP per head	$43,005
GDP	AED1,521bn	GDP per head in purchasing	
Av. ann. growth in real		power parity (USA=100)	119.6
GDP 2013–18	2.9%	Economic freedom index	76.2

Origins of GDP		**Components of GDP**	
	% of total		% of total
Agriculture	0.7	Private consumption	38.5
Industry, of which:	46.8	Public consumption	13.1
manufacturing	8.9	Investment	22.4
Services	52.5	Exports	93.9
		Imports	-68.0

Structure of employment

	% of total		% of labour force
Agriculture	1.4	Unemployed 2019	2.2
Industry	34.4	Av. ann. rate 2009–19	2.3
Services	64.2		

Energy

	m TOE		
Total output	249.4	Net energy imports as %	
Total consumption	117.7	of energy use	-112
Consumption per head			
kg oil equivalent	12,521		

Inflation and finance

		% change 2018–19	
Consumer price			
inflation 2019	-1.9%	Narrow money (M1)	7.9
Av. ann. inflation 2014–19	1.7%	Broad money	8.0
Central bank policy rate, Dec. 2019	2.50%		

Exchange rates

	end 2019		December 2019
			2010 = 100
AED per $	3.67	Effective rates	142.3
AED per SDR	5.08	– nominal	142.3
AED per €	4.12	– real	...

Trade

Principal exports		Principal imports	
	$bn fob		*$bn cif*
Re-exports	141.9	Precious stones & metals	51.0
Crude oil	32.0	Machinery & electrical equip.	42.6
Petroleum products	25.9	Vehicles & other transport	
Gas	9.4	equipment	29.8
		Base metals & related products	15.0
Total incl. others	**321.0**	Total incl. others	**261.5**

Main export destinations		Main origins of imports	
	% of total		*% of total*
India	10.1	China	15.5
Japan	9.8	India	9.4
China	6.0	United States	8.5
Oman	4.3	Japan	5.6

Balance of payments, reserves and debt, $bn

Visible exports fob	321.0	Change in reserves	4.1
Visible imports fob	-235.4	Level of reserves	
Trade balance	85.7	end Dec.	99.5
Invisibles inflows	81.4	No. months of import cover	3.8
Invisibles outflows	-80.5	Official gold holdings, m oz	0.2
Net transfers	-46.1	Foreign debt	238.1
Current account balance	40.5	– as % of GDP	57.5
– as % of GDP	9.8	– as % of total exports	58.0
Capital balance	-34.6	Debt service ratio	7.3
Overall balance	3.5		

Health and education

Health spending, % of GDP	3.3	Education spending, % of GDP	...
Doctors per 1,000 pop.	2.5	Enrolment, %: primary	108
Hospital beds per 1,000 pop.	1.2	secondary	105
At least basic drinking water,		tertiary	...
% of pop.	98.0		

Society

No. of households, m	1.7	Cost of living, Dec. 2019	
Av. no. per household	5.6	New York = 100	73
Marriages per 1,000 pop.	...	Cars per 1,000 pop.	226
Divorces per 1,000 pop.	...	Telephone lines per 100 pop.	24.3
Religion, % of pop.		Mobile telephone subscribers	
Muslim	76.9	per 100 pop.	208.5
Christian	12.6	Internet access, %	98.5
Hindu	6.6	Broadband subs per 100 pop.	31.4
Other	2.8	Broadband speed, Mbps	4.4
Non-religious	1.1		
Jewish	<0.1		

UNITED KINGDOM

Area, sq km	243,610	Capital	London
Arable as % of total land	24.9	Currency	Pound (£)

People

Population, m	67.1	Life expectancy: men	80.2 yrs
Pop. per sq km	276.7	women	83.3 yrs
Average annual rate of change		Adult literacy	...
in pop. 2015–20, %	0.6	Fertility rate (per woman)	1.8
Pop. aged 0–19, 2020, %	23.1	Urban population, %	83.4
Pop. aged 65 and over, 2020, %	18.7		per 1,000 pop.
No. of men per 100 women	97.7	Crude birth rate	11.5
Human Development Index	92.0	Crude death rate	9.5

The economy

GDP	$2,855bn	GDP per head	$42,944
GDP	£2,118bn	GDP per head in purchasing	
Av. ann. growth in real		power parity (USA=100)	73.2
GDP 2013–18	2.0%	Economic freedom index	79.3

Origins of GDP		**Components of GDP**	
	% of total		% of total
Agriculture	0.7	Private consumption	65.5
Industry, of which:	20.3	Public consumption	18.5
manufacturing	6.3	Investment	17.3
Services	79.0	Exports	30.0
		Imports	-31.8

Structure of employment

	% of total		% of labour force
Agriculture	1.0	Unemployed 2019	4.0
Industry	17.9	Av. ann. rate 2009–19	3.9
Services	81.1		

Energy

	m TOE		
Total output	132.8	Net energy imports as %	
Total consumption	207.5	of energy use	36
Consumption per head			
kg oil equivalent	3,135		

Inflation and finance

Consumer price			% change 2018–19
inflation 2019	1.8%	Narrow money (M1)	-1.6
Av. ann. inflation 2014–19	1.5%	Broad money	-1.3
Central bank policy rate, Dec. 2019	0.75%		

Exchange rates

	end 2019		December 2019
£ per $	0.76	Effective rates	2010 = 100
£ per SDR	1.05	– nominal	101.0
£ per €	0.85	– real	101.1

Trade

Principal exports

	$bn fob
Machinery & transport equip.	178.6
Chemicals & related products	73.3
Mineral fuels & lubricants	47.7
Food, drink & tobacco	30.2
Total incl. others	**466.4**

Principal imports

	$bn cif
Machinery & transport equip.	243.5
Chemicals & related products	77.9
Mineral fuels & lubricants	67.3
Food, drink & tobacco	62.1
Total incl. others	**652.4**

Main export destinations

	% of total
United States	14.0
Germany	10.0
Netherlands	7.4
France	6.8
EU28	47.1

Main origins of imports

	% of total
Germany	14.0
United States	10.0
China	9.6
Netherlands	8.6
EU28	52.8

Balance of payments, reserves and aid, $bn

Visible exports fob	466.6	Overall balance	24.3
Visible imports fob	-652.4	Change in reserves	21.8
Trade balance	-185.8	Level of reserves	
Invisibles inflows	700.8	end Dec.	172.6
Invisibles outflows	-591.1	No. months of import cover	1.7
Net transfers	-34.1	Official gold holdings, m oz	10.0
Current account balance	-110.2	Aid given	19.5
– as % of GDP	-3.9	– as % of GNI	0.7
Capital balance	131.8		

Health and education

Health spending, % of GDP	9.6	Education spending, % of GDP	5.5
Doctors per 1,000 pop.	2.8	Enrolment, %: primary	101
Hospital beds per 1,000 pop.	2.8	secondary	126
At least basic drinking water,		tertiary	60
% of pop.	100		

Society

No. of households, m	29.0	Cost of living, Dec. 2019	
Av. no. per household	2.3	New York = 100	80
Marriages per 1,000 pop.	4.4	Cars per 1,000 pop.	505
Divorces per 1,000 pop.	1.8	Telephone lines per 100 pop.	47.5
Religion, % of pop.		Mobile telephone subscribers	
Christian	71.1	per 100 pop.	118.4
Non-religious	21.3	Internet access, %	94.9
Muslim	4.4	Broadband subs per 100 pop.	39.6
Other	1.4	Broadband speed, Mbps	18.6
Hindu	1.3		
Jewish	0.5		

UNITED STATES

Area, sq km	9,831,510	Capital	Washington DC
Arable as % of total land	16.6	Currency	US dollar ($)

People

Population, m	327.1	Life expectancy: men	76.6 yrs
Pop. per sq km	33.3	women	81.7 yrs
Average annual rate of change		Adult literacy	…
in pop. 2015–20, %	0.6	Fertility rate (per woman)	1.8
Pop. aged 0–19, 2020, %	24.8	Urban population, %	82.3
Pop. aged 65 and over, 2020, %	16.6		per 1,000 pop.
No. of men per 100 women	97.9	Crude birth rate	12.0
Human Development Index	92.0	Crude death rate	9.2

The economy

GDP	$20,544bn	GDP per head	$62,795
Av. ann. growth in real		GDP per head in purchasing	
GDP 2013–18	2.4%	power parity (USA=100)	100.0
		Economic freedom index	76.6

Origins of GDP		**Components of GDP**	
	% of total		% of total
Agriculture	0.8	Private consumption	68.1
Industry, of which:	18.6	Public consumption	14.1
manufacturing	11.4	Investment	21.0
Services	80.6	Exports	12.2
		Imports	-15.3

Structure of employment

	% of total		% of labour force
Agriculture	1.3	Unemployed 2019	3.9
Industry	19.8	Av. ann. rate 2009-19	3.7
Services	78.9		

Energy

	m TOE		
Total output	2,223.0	Net energy imports as %	
Total consumption	2,463.9	of energy use	10
Consumption per head			
kg oil equivalent	7,594		

Inflation and finance

			% change 2018–19
Consumer price			
inflation 2019	1.8%	Narrow money (M1)	2.9
Av. ann. inflation 2014–19	1.6%	Broad money	8.4
Money market rate, Dec. 2019	1.64%		

Exchange rates

	end 2019		December 2019
$ per SDR	1.38	Effective rates	2010 = 100
$ per €	1.12	– nominal	121.9
		– real	116.3

Trade

Principal exports	$bn fob	Principal imports	$bn fob
Capital goods, excl. vehicles	562.9	Capital goods, excl. vehicles	692.6
Industrial supplies	541.7	Consumer goods, excl. vehicles	646.8
Consumer goods, excl. vehicles	206.0	Industrial supplies	575.6
Vehicles & products	158.8	Vehicles & products	372.2
Total incl. others	**1,666.0**	**Total incl. others**	**2,540.8**

Main export destinations	% of total	Main origins of imports	% of total
Canada	18.0	China	21.2
Mexico	15.9	Mexico	13.6
China	7.2	Canada	12.5
Japan	4.5	Japan	5.6
EU28	19.2	EU28	19.2

Balance of payments, reserves and aid, $bn

Visible exports fob	1,674.3	Overall balance	5.0
Visible imports fob	-2,561.7	Change in reserves	-0.6
Trade balance	-887.3	Level of reserves	
Invisibles inflows	1,911.2	end Dec.	449.2
Invisibles outflows	-1,397.5	No. months of import cover	1.4
Net transfers	-117.3	Official gold holdings, m oz	261.5
Current account balance	-491.0	Aid given	33.8
– as % of GDP	-2.4	– as % of GNI	0.2
Capital balance	453.7		

Health and education

Health spending, % of GDP	17.1	Education spending, % of GDP	...
Doctors per 1,000 pop.	2.6	Enrolment, %: primary	102
Hospital beds per 1,000 pop.	2.9	secondary	99
At least basic drinking water,		tertiary	88
% of pop.	99.3		

Society

No. of households, m	138.5	Cost of living, Dec. 2019	
Av. no. per household	2.4	New York = 100	100
Marriages per 1,000 pop.	6.9	Cars per 1,000 pop.	365
Divorces per 1,000 pop.	2.5	Telephone lines per 100 pop.	33.6
Religion, % of pop.		Mobile telephone subscribers	
Christian	78.3	per 100 pop.	129.0
Non-religious	16.4	Internet access, %	87.3
Other	2.0	Broadband subs per 100 pop.	33.8
Jewish	1.8	Broadband speed, Mbps	25.9
Muslim	0.9		
Hindu	0.6		

VENEZUELA

Area, sq km	912,050	Capital	Caracas
Arable as % of total land	3.1	Currency	Bolivar (Bs)

People

Population, m	28.9	Life expectancy: men	68.6 yrs
Pop. per sq km	31.7	women	76.3 yrs
Average annual rate of change		Adult literacy	97.1
in pop. 2015–20, %	-1.1	Fertility rate (per woman)	2.3
Pop. aged 0–19, 2020, %	36.1	Urban population, %	88.2
Pop. aged 65 and over, 2020, %	8.0		per 1,000 pop.
No. of men per 100 women	96.8	Crude birth rate	18.0
Human Development Index	72.6	Crude death rate	7.3

The economy

GDP	$98bn	GDP per head	$3,411
GDP	Bs2,039bn	GDP per head in purchasing	
Av. ann. growth in real		power parity (USA=100)	17.2
GDP 2013–18	-12.7%	Economic freedom index	25.2

Origins of GDP[a]		Components of GDP[a]	
	% of total		% of total
Agriculture	2.9	Private consumption	100.5
Industry, of which:	28.4	Public consumption	9.0
manufacturing	...	Investment	-3.7
Services	68.7	Exports	30.7
		Imports	-36.4

Structure of employment

	% of total		% of labour force
Agriculture	8.3	Unemployed 2019	7.2
Industry	16.6	Av. ann. rate 2009-19	8.8
Services	75.1		

Energy

	m TOE		
Total output	157.4	Net energy imports as %	
Total consumption	62.9	of energy use	-150
Consumption per head			
kg oil equivalent	1,966		

Inflation and finance

		% change 2018–19	
Consumer price			
inflation 2019	19,906%	Narrow money (M1)	1,129
Av. ann. inflation 2014–19	2,133%	Broad money	4,946
Prime bank loan rate, Dec. 2019	21.69%		

Exchange rates

	end 2019		December 2019
Bs per $	...	Effective rates	2010 = 100
Bs per SDR	...	– nominal	...
Bs per €	...	– real	...

Trade

Principal exports		**Principal imports**	
	$bn fob		*$bn cif*
Oil	29.8	Consumer goods	9.3
Non-oil	3.2	Capital goods	1.9
		Intermediate goods	1.6
Total incl. others	**33.7**	Total incl. others	**13.8**

Main export destinations		**Main origins of imports**	
	% of total		*% of total*
United States	33.8	United States	31.3
India	19.4	China	10.4
China	14.7	Mexico	9.6
Netherlands Antilles	7.7	Brazil	4.1

Balance of payments, reserves and debt, $bn

Visible exports fob	33.7	Change in reserves	-1.2
Visible imports fob	-12.8	Level of reserves	
Trade balance	20.9	end Dec.	8.6
Invisibles inflows	1.6	No. months of import cover	3.6
Invisibles outflows	-15.8	Official gold holdings, m oz	5.2
Net transfers	2.0	Foreign debt	154.9
Current account balance	8.6	– as % of GDP	63.9
– as % of GDP	8.7	– as % of total exports	433.6
Capital balance	3.8	Debt service ratio	27.9
Overall balance	8.5		

Health and education

Health spending, % of GDP	1.2	Education spending, % of GDP	...
Doctors per 1,000 pop.	...	Enrolment, %: primary	97
Hospital beds per 1,000 pop.	0.8	secondary	88
At least basic drinking water,		tertiary	...
% of pop.	95.7		

Society

No. of households, m	8.6	Cost of living, Dec. 2019	
Av. no. per household	3.4	New York = 100	36
Marriages per 1,000 pop.	2.6	Cars per 1,000 pop.	127
Divorces per 1,000 pop.	0.7	Telephone lines per 100 pop.	19.2
Religion, % of pop.		Mobile telephone subscribers	
Christian	89.3	per 100 pop.	71.8
Non-religious	10.0	Internet access, %	72.0
Muslim	0.3	Broadband subs per 100 pop.	9.0
Other	0.3	Broadband speed, Mbps	1.2
Hindu	<0.1		
Jewish	<0.1		

VIETNAM

Area, sq km	331,230	Capital	Hanoi
Arable as % of total land	22.6	Currency	Dong (D)

People

Population, m	95.5	Life expectancy: men	71.7 yrs
Pop. per sq km	288.5	women	79.9 yrs
Average annual rate of change		Adult literacy	95.0
in pop. 2015–20, %	1.0	Fertility rate (per woman)	2.1
Pop. aged 0–19, 2020, %	29.9	Urban population, %	35.9
Pop. aged 65 and over, 2020, %	7.9		per 1,000 pop.
No. of men per 100 women	99.7	Crude birth rate	16.9
Human Development Index	69.3	Crude death rate	6.6

The economy

GDP	$245bn	GDP per head	$2,567
GDP	D5,535trn	GDP per head in purchasing	
Av. ann. growth in real		power parity (USA=100)	11.9
GDP 2013–18	6.6%	Economic freedom index	58.8

Origins of GDP		**Components of GDP**	
	% of total		% of total
Agriculture	16.3	Private consumption	67.6
Industry, of which:	38.0	Public consumption	6.5
manufacturing	17.8	Investment	26.5
Services	45.7	Exports	105.8
		Imports	-102.5

Structure of employment

	% of total		% of labour force
Agriculture	37.4	Unemployed 2019	2.0
Industry	27.6	Av. ann. rate 2009-19	2.0
Services	35.0		

Energy

	m TOE		
Total output	64.5	Net energy imports as %	
Total consumption	81.8	of energy use	21
Consumption per head			
kg oil equivalent	856		

Inflation and finance

			% change 2018–19
Consumer price			
inflation 2019	2.8%	Narrow money (M1)	24.0
Av. ann. inflation 2014–19	2.6%	Broad money	13.6
Deposit rate, Dec. 2019	4.90%		

Exchange rates

	end 2019		December 2019
D per $	23,155.0	Effective rates	2010 = 100
D per SDR	32,019.3	– nominal	...
D per €	26,016.9	– real	...

Trade

Principal exports		Principal imports	
	$bn fob		*$bn cif*
Telephones & mobile phones	49.2	Electronics, computers & parts	43.1
Textiles & garments	30.5	Machinery & equipment	32.9
Computers & electronic products	29.6	Telephones & mobile phones	15.9
Footwear	16.2	Textiles	12.8
Total incl. others	**243.7**	Total incl. others	**236.9**

Main export destinations		Main origins of imports	
	% of total		*% of total*
United States	19.9	China	28.3
China	17.3	South Korea	20.5
Japan	7.9	Japan	8.2
South Korea	7.6	Taiwan	5.7

Balance of payments, reserves and debt, $bn

Visible exports fob	243.7	Change in reserves	6.4
Visible imports fob	-227.2	Level of reserves	
Trade balance	16.5	end Dec.	55.5
Invisibles inflows	16.4	No. months of import cover	2.5
Invisibles outflows	-35.9	Official gold holdings, m oz	0.0
Net transfers	8.9	Foreign debt	108.1
Current account balance	5.9	– as % of GDP	35.7
– as % of GDP	2.4	– as % of total exports	39.2
Capital balance	8.5	Debt service ratio	6.7
Overall balance	6.0		

Health and education

Health spending, % of GDP	5.5	Education spending, % of GDP	4.2
Doctors per 1,000 pop.	0.8	Enrolment, %: primary	111
Hospital beds per 1,000 pop.	2.6	secondary	...
At least basic drinking water,		tertiary	29
% of pop.	94.7		

Society

No. of households, m	28.9	Cost of living, Dec. 2019	
Av. no. per household	3.3	New York = 100	70
Marriages per 1,000 pop.	...	Cars per 1,000 pop.	23
Divorces per 1,000 pop.	...	Telephone lines per 100 pop.	4.5
Religion, % of pop.		Mobile telephone subscribers	
Other	45.6	per 100 pop.	147.2
Non-religious	29.6	Internet access, %	70.4
Buddhist	16.4	Broadband subs per 100 pop.	13.6
Christian	8.2	Broadband speed, Mbps	6.7
Muslim	0.2		
Jewish	<0.1		

ZIMBABWE

Area, sq km	390,760	Capital	Harare
Arable as % of total land	10.3	Currency	Zimbabwe dollar (Z$)

People

Population, m	14.4	Life expectancy: men	60.4 yrs
Pop. per sq km	36.9	women	63.7 yrs
Average annual rate of change		Adult literacy	88.7
in pop. 2015–20, %	1.5	Fertility rate (per woman)	3.6
Pop. aged 0–19, 2020, %	52.9	Urban population, %	32.2
Pop. aged 65 and over, 2020, %	3.0		per 1,000 pop.
No. of men per 100 women	91.3	Crude birth rate	30.8
Human Development Index	56.3	Crude death rate	7.6

The economy

GDP	$31bn	GDP per head	$2,147
GDP	Z$43bn	GDP per head in purchasing	
Av. ann. growth in real		power parity (USA=100)	4.8
GDP 2013–18	3.1%	Economic freedom index	43.1

Origins of GDP		Components of GDP	
	% of total		% of total
Agriculture	9.2	Private consumption	74.2
Industry, of which:	22.9	Public consumption	15.9
manufacturing	11.7	Investment	12.6
Services	67.4	Exports	22.9
		Imports	-25.5

Structure of employment

	% of total		% of labour force
Agriculture	66.5	Unemployed 2019	5.1
Industry	6.6	Av. ann. rate 2009-19	5.0
Services	26.9		

Energy

	m TOE		
Total output	2.8	Net energy imports as %	
Total consumption	4.0	of energy use	31
Consumption per head			
kg oil equivalent	243		

Inflation and finance

			% change 2018–19
Consumer price			
inflation 2019	255.3%	Narrow money (M1)	217.0
Av. ann. inflation 2014–19	30.7%	Broad money	249.8
Deposit rate, Dec. 2019	3.39%		

Exchange rates

	end 2019		December 2019
Z$ per $	16.77	Effective rates	2010 = 100
Z$ per SDR	23.19	– nominal	...
Z$ per €	18.84	– real	...

Trade

Principal exports[a]

	$bn fob
Gold	0.9
Tobacco	0.9
Ferro-alloys	0.7
Platinum	0.1
Total incl. others	**3.3**

Principal imports[a]

	$bn cif
Fuels & mining products	0.4
Machinery & transport equip.	0.4
Manufactures	0.3
Chemicals	0.2
Total	**4.1**

Main export destinations

	% of total
South Africa	60.3
United Arab Emirates	21.3
Mozambique	11.4
Zambia	1.9

Main origins of imports

	% of total
South Africa	51.0
Zambia	20.3
China	3.1
United States	3.0

Balance of payments[b], reserves and debt, $bn

Visible exports fob	4.3	Change in reserves	-0.2
Visible imports fob	-5.5	Level of reserves	
Trade balance	-1.2	end Dec.	0.1
Invisibles inflows	0.6	No. months of import cover	0.2
Invisibles outflows	-1.5	Official gold holdings, m oz	0.0
Net transfers	1.7	Foreign debt	12.3
Current account balance	-0.3	– as % of GDP	39.6
– as % of GDP	-1.3	– as % of total exports	184.4
Capital balance	0.1	Debt service ratio	13.8
Overall balance	-0.6		

Health and education

Health spending, % of GDP	6.6	Education spending, % of GDP	4.6
Doctors per 1,000 pop.	0.2	Enrolment, %: primary	...
Hospital beds per 1,000 pop.	1.7	secondary	...
At least basic drinking water,		tertiary	...
% of pop.	64.1		

Society

No. of households, m	3.1	Cost of living, Dec. 2019	
Av. no. per household	4.6	New York = 100	...
Marriages per 1,000 pop.	...	Cars per 1,000 pop.	60
Divorces per 1,000 pop.	...	Telephone lines per 100 pop.	1.9
Religion, % of pop.		Mobile telephone subscribers	
Christian	87.0	per 100 pop.	89.4
Non-religious	7.9	Internet access, %	27.1
Other	4.2	Broadband subs per 100 pop.	1.4
Muslim	0.9	Broadband speed, Mbps	2.9
Hindu	<0.1		
Jewish	<0.1		

a 2016   b 2017

EURO AREA[a]

Area, sq km	2,759,788	Capital	–
Arable as % of total land	23.9	Currency	Euro (€)

People

Population, m	339.8	Life expectancy: men	80.3 yrs
Pop. per sq km	130.1	women	85.2 yrs
Average annual rate of change		Adult literacy	...
in pop. 2015–20, %	0.2	Fertility rate (per woman)	1.5
Pop. aged 0–19, 2020, %	20.1	Urban population, %	76.9
Pop. aged 65 and over, 2020, %	21.2		per 1,000 pop.
No. of men per 100 women	95.9	Crude birth rate	9.4
Human Development Index	90.4	Crude death rate	10.5

The economy

GDP	$13.6trn	GDP per head	$46,759
GDP	€11.8trn	GDP per head in purchasing	
Av. ann. growth in real		power parity (USA=100)	74.5
GDP 2013–18	2.0%	Economic freedom index	68.8

Origins of GDP		**Components of GDP**	
	% of total		% of total
Agriculture	1.7	Private consumption	53.7
Industry, of which:	24.9	Public consumption	20.4
manufacturing	16.9	Investment	21.6
Services	73.4	Exports	48.0
		Imports	-43.7

Structure of employment

	% of total		% of labour force
Agriculture	3.0	Unemployed 2019	8.2
Industry	23.3	Av. ann. rate 2009-19	7.5
Services	73.7		

Energy

	m TOE		
Total output	427.6	Net energy imports as %	
Total consumption	1,285.0	of energy use	67
Consumption per head			
kg oil equivalent	3,809		

Inflation and finance

			% change 2018–19
Consumer price			
inflation 2019	1.2%	Narrow money (M1)	7.6
Av. ann. inflation 2014–19	1.0%	Broad money	5.2
Deposit rate, Dec. 2019	0.21%		

Exchange rates

	end 2019		December 2019
			2010 = 100
€ per $	0.78	Effective rates	
€ per SDR	1.23	– nominal	105.8
		– real	91.7

Trade[b]

Principal exports

	$bn fob
Machinery & transport equip.	957.7
Other manufactured goods	518.6
Chemicals & related products	419.6
Food, drink & tobacco	143.8
Mineral fuels & lubricants	135.7
Total incl. others	**2,312.5**

Principal imports

	$bn cif
Machinery & transport equip.	732.6
Other manufactured goods	584.6
Mineral fuels & lubricants	486.1
Chemicals & related products	241.0
Food, drink & tobacco	133.2
Total incl. others	**2,338.7**

Main export destinations

	% of total
United States	20.8
China	10.8
Switzerland	8.0
Russia	4.3
Turkey	3.9

Main origins of imports

	% of total
China	20.0
United States	13.6
Russia	8.5
Switzerland	5.5
Norway	4.2

Balance of payments, reserves and aid, $bn

Visible exports fob	2,767.3	Overall balance	38.2
Visible imports fob	-2,412.4	Change in reserves	20.8
Trade balance	354.9	Level of reserves	
Invisibles inflows	2,012.1	end Dec.	821.6
Invisibles outflows	-1,763.4	No. months of import cover	2.4
Net transfers	-178.3	Official gold holdings, m oz	346.5
Current account balance	425.2	Aid given	58.7
– as % of GDP	3.1	– as % of GDP	0.4
Capital balance	-442.5		

Health and education

Health spending, % of GDP	10.1	Education spending, % of GDP	4.8
Doctors per 1,000 pop.	3.9	Enrolment, %: primary	103
Hospital beds per 1,000 pop.	6.2	secondary	108
At least basic drinking water,		tertiary	74
% of pop.	99.8		

Society

No. of households, m, m	150.3	Telephone lines per 100 pop.	34.5
Av. no. per household	2.2	Mobile telephone subscribers	
Marriages per 1,000 pop.	4.0	per 100 pop.	131.6
Divorces per 1,000 pop.	1.8	Internet access, %	81.3
Cost of living, Dec. 2019		Broadband subs per 100 pop.	34.6
New York = 100	...	Broadband speed, Mbps	...
Cars per 1,000 pop.	539		

a Data generally refer to the 19 EU members that had adopted the euro as at December 31 2017: Austria, Belgium, Cyprus, Estonia, Finland, France, Germany, Greece, Ireland, Italy, Latvia, Lithuania, Luxembourg, Malta, Netherlands, Portugal, Slovakia, Slovenia and Spain.

b EU28, excluding intra-trade.

WORLD

Area, sq km	132,025,199	Capital	...
Arable as % of total land	11.1	Currency	...

People

Population, m	7,631.1	Life expectancy: men	71.7 yrs
Pop. per sq km	51.3	women	76.4 yrs
Average annual rate of change		Adult literacy	86.3
in pop. 2015–20, %	1.1	Fertility rate (per woman)	2.5
Pop. aged 0–19, 2020, %	33.3	Urban population, %	55.3
Pop. aged 65 and over, 2020, %	9.3		per 1,000 pop.
No. of men per 100 women	101.7	Crude birth rate	18.5
Human Development Index	73.1	Crude death rate	7.7

The economy

GDP	$84.9trn	GDP per head	$17,948
Av. ann. growth in real		GDP per head in purchasing	
GDP 2013–18	3.6%	power parity (USA=100)	28.6
		Economic freedom index	59.9

Origins of GDP		**Components of GDP**	
	% of total		% of total
Agriculture	4.7	Private consumption	57.7
Industry, of which:	30.8	Public consumption	16.9
manufacturing	18.7	Investment	24.4
Services	65.0	Exports	30.1
		Imports	-29.3

Structure of employment

	% of total		% of labour force
Agriculture	26.9	Unemployed 2019	5.4
Industry	23.0	Av. ann. rate 2009–19	5.4
Services	50.1		

Energy

	m TOE		
Total output	14,518.0	Net energy imports as %	
Total consumption	14,712.1	of energy use	...
Consumption per head			
kg oil equivalent	1,948		

Inflation and finance

Consumer price			% change 2018–19
inflation 2019	3.6%	Narrow money (M1)	5.1
Av. ann. inflation 2014–19	3.2%	Broad money	4.9
LIBOR $ rate, 3-month, Dec. 2019	1.91%		

Trade

World exports

	$bn fob		$bn fob
Manufactures	13,485	Ores & minerals	784
Food	2,470	Agricultural raw materials	294
Fuels	1,666		
		Total incl. others	**19,600**

Main export destinations		**Main origins of imports**	
	% of total		% of total
United States	12.8	China	13.5
China	9.7	United States	8.4
Germany	6.6	Germany	7.8
France	3.6	Japan	4.0
United Kingdom	3.6	South Korea	3.3

Balance of payments, reserves and aid, $bn

Visible exports fob	19,083.9	Capital balance	-155.4
Visible imports fob	-18,827.5	Overall balance	0.0
Trade balance	256.4	Change in reserves	-10.8
Invisibles inflows	10,420.0	Level of reserves	
Invisibles outflows	-10,205.8	end Dec.	13,203.6
Net transfers	-118.7	No. months of import cover	5.5
Current account balance	352.0	Official gold holdings, m oz	1,100.0
– as % of GDP	0.4		

Health and education

Health spending, % of GDP	...	Education spending, % of GDP	4.1
Doctors per 1,000 pop.	1.6	Enrolment, %: primary	104
Hospital beds per 1,000 pop.	2.7	secondary	76
At least basic drinking water,		tertiary	38
% of pop.	89.6		

Society

No. of households, m, m	2,103.4	Cost of living, Dec. 2019	
Av. no. per household	3.6	New York = 100	...
Marriages per 1,000 pop.	...	Cars per 1,000 pop.	132
Divorces per 1,000 pop.	...	Telephone lines per 100 pop.	17.4
Religion, % of pop.		Mobile telephone subscribers	
Christian	31.5	per 100 pop.	108.2
Muslim	23.2	Internet access, %	49.1
Non-religious	16.3	Broadband subs per 100 pop.	16.2
Hindu	15.0	Broadband speed, Mbps	...
Other	13.8		
Jewish	0.2		

WORLD RANKINGS QUIZ

Test your knowledge with our world rankings quiz. Answers can be found on the pages indicated.

Geography and demographics

1 Which is the world's second largest country?
a The United States b Canada c Brazil d China *page 12*

2 Which is the largest country in Africa?
a Congo-Kinshasa b Sudan c Algeria d Libya *page 12*

3 Which country has the longer coastline?
a Brazil b United Kingdom *page 12*

4 In which country are eight of the world's ten highest mountains?
a Pakistan b China c Nepal *page 13*

5 What is the world's longest river?
a Amazon b Nile c Yangtze (Chang Jiang) *page 13*

6 In which country would you find the Great Basin?
a Australia b United States c China *page 13*

7 Which is the largest of the Great Lakes?
a Superior b Erie c Huron *page 13*

8 Which of these has the biggest population?
a Bangladesh b Brazil c Nigeria d Pakistan *page 14*

9 Which country has the slowest rate of population growth?
a Bulgaria b Puerto Rico c Syria d Venezuela *page 15*

10 The ten countries with highest fertility rates are all in Africa.
a True b False *page 17*

11 Which country has the lowest median age?
a Angola b Niger c Uganda d Zambia *page 18*

12 Shanghai has a bigger population than Delhi.
a True b False *page 19*

13 Which city has the slowest rate of population growth?
a Beirut b Bucharest c Naples d Tokyo *page 19*

14 Which of these has the largest foreign-born population?
a France b Italy c Spain d United Kingdom *page 21*

Business and economics

1 Which country has the highest GDP per person in dollars?
a United States b Germany c Ireland d Qatar   *page 26*

2 Which country scores highest in the world on the UN Human Development Index?
a Canada b Norway c Switzerland d New Zealand *page 28*

3 Which country has the most equal distribution of household income, according to the Gini coefficient measure?
a Ukraine b Belarus c Slovenia d Moldova   *page 28*

4 Which country has the most millionaires?
a China b Japan c United Kingdom d United States *page 29*

5 The fastest-growing economy over the last decade is on which continent?
a Africa b Asia c Europe d South America   *page 30*

6 Which country is most dependent on foreign-trade?
a United States b Cambodia c Vietnam d South Sudan
page 32

7 Which of these countries recorded a current-account surplus in 2018?
a Brazil b United States c India d Russia   *page 34*

8 Which country receives the highest level of remittances from workers in foreign countries?
a Mexico b India c France d Philippines   *page 36*

9 Which country holds the world's largest official gold reserves?
a Germany b China c United States d Russia   *page 36*

10 In which of these countries is a Big Mac most expensive?
a Canada b Denmark c France d Sweden   *page 37*

11 Which country had an inflation rate of almost 20,000% in 2019?
a Sudan b Argentina c Zimbabwe d Venezuela   *page 38*

12 Which of these has the highest foreign debt?
a Brazil b Mexico c Russia d Turkey   *page 40*

13 Which of these received the most foreign aid?
   a Bangladesh b Ethiopia c Iraq d West Bank & Gaza
page 42

14 Which country provides the most foreign aid?
   a Norway b Saudi Arabia c United Kingdom
   d United States
page 43

15 Which country has the highest manufacturing output?
   a Italy b France c United Kingdom d Spain *page 45*

16 Japan has a larger agricultural output than France.
   a True b False
page 46

17 Macau and Singapore have no economic dependence on
   agriculture. Which country has the most?
   a Chad b Benin c Sierra Leone d Sudan *page 46*

18 Argentina produces more cereal than Russia.
   a True b False
page 47

19 Which country is the largest consumer of coffee?
   a Canada b Russia c Ethiopia d Japan *page 49*

20 Which country is the most energy efficient?
   a Macau b Chad c France d Nigeria *page 54*

21 Which OECD country has highest % of women on
   company boards?
   a Iceland b France c Sweden d New Zealand *page 56*

22 Greece has a lower rate of youth unemployment than
   Spain.
   a True b False
page 57

23 Which country has the most unpaid work on average?
   aPortugal b France c Mexico d Poland *page 58*

24 Which country has the highest foreign direct investment
   inflows?
   a China b Switzerland c Brazil d United States *page 59*

25 Which country has the highest "brain-gain" (attracting the
   highly skilled from abroad) score?
   a Switzerland b United States c Singapore d Luxembourg
page 60

Politics and society

1   Guinea has the lowest adult literacy rate.
    a True **b** False   *page 69*

2   What is the mean age of Bangladeshi women at their first
    marriage?
    a 15.9 **b** 16.5 **c** 17.7 **d** 18.8   *page 71*

3   Which country has the biggest average household size?
    a Pakistan **b** Yemen **c** Senegal **d** Niger   *page 72*

4   Which country gives most generously?
    a Singapore **b** United States **c** Myanmar **d** Sri Lanka *page 73*

5   Which country has the most vehicles relative to its road
    network?
    a Monaco **b** Hong Kong **c** United Arab Emirates **d** Kuwait
                                                                page 75

6   Which airport is the busiest for passenger traffic?
    a Atlanta **b** Frankfurt **c** Beijing **d** Dallas *page 78*

7   Which country transports the most freight by rail?
    a China **b** India **c** United States **d** Russia *page 79*

8   Which country has the most registered ships?
    a China **b** Indonesia **c** Japan **d** Panama   *page 80*

9   Which country has the largest prison population?
    a Russia **b** India **c** Brazil **d** Thailand   *page 81*

10 The US spends more on defence than the next ten highest
    spenders combined.
    a True **b** False   *page 82*

11 Which country spends the most on defence per person?
    a United States **b** Israel **c** Oman **d** Saudi Arabia *page 82*

12 Which country exported the highest value of arms in 2019?
    a Germany **b** United Kingdom **c** France   *page 83*

13 Which country suffered the most terrorist attacks in 2018?
    a India **b** Iraq **c** Nigeria **d** Afghanistan   *page 83*

14 Which of these countries has most satellites in space?
    a France **b** Germany **c** United Kingdom **d** Italy *page 84*

Health and welfare

1   Which country's men live the longest?
    a Iceland **b** Japan **c** Monaco **d** Sweden   *page 90*

2   Lesotho has the lowest life expectancy in the world.
    a True **b** False   *page 91*

3   Men in Zimbabwe live longer than women in Nigeria.
    a True **b** False   *page 91*

4   Which country has the lowest death rate?
    a Germany **b** Japan **c** Nigeria **d** Russia   *page 92*

5   Pakistan has a higher infant mortality rate than Nigeria.
    a True **b** False   *page 93*

6   Which country has the highest rate of deaths from cancer?
    a Netherlands **b** Poland **c** Russia **d** United Kingdom
    page 94

7   Which country has the highest rate of death from Ischaemic
    heart disease?
    a Latvia **b** Ukraine **c** Belarus **d** Russia   *page 94*

8   Which country has the lowest % of one-year-olds
    immunised against measles?
    a Angola **b** Afghanistan **c** Bolivia **d** Dominican Republic
    page 95

9   The US has the biggest health spending as a % of its GDP.
    Who is second?
    a Switzerland **b** Sierra Leone **c** Cuba **d** Malawi   *page 96*

10 Which country has the highest out-of-pocket health
    spending per person?
    a United States **b** Singapore **c** Switzerland **d** Austria
    page 96

11 Child obesity is highest in which country?
    a Kuwait **b** Qatar **c** Bahrain **d** United States   *page 97*

12 Which country has the highest proportion of underweight
    children?
    a Vietnam **b** India **c** Nepal **d** Congo-Kinshasa   *page 97*

Culture and entertainment

1 Taiwan has more landline telephones per person than the United Kingdom.
   a True **b** False   *page 98*

2 Which country has the fastest broadband?
   a United States **b** Sweden **c** Taiwan **d** Singapore *page 99*

3 Iceland published the most books per person in 2018.
   a True **b** False   *page 100*

4 Which country pays the most for cinema tickets?
   a Bahrain **b** Iceland **c** Japan **d** Sweden   *page 101*

5 Which country has the freest press?
   a Costa Rica **b** Jamaica **c** Netherlands **d** Norway *page 102*

6 China imprisons more journalists than any other country.
   a True **b** False   *page 102*

7 Which country has produced most Nobel prize winners in Literature?
   a Germany **b** Italy **c** France **d** United Kingdom *page 103*

8 Which country was most successful at summer Olympics?
   a Soviet/Unified team **b** East Germany **c** West Germany
                                                             page 105

9 France had more doping violations than Italy in 2017.
   a True **b** False   *page 105*

10 Which of these drinks the most beer per person?
    a Austria **b** Germany **c** Ireland **d** Spain   *page 106*

11 Which of these consumes the most wine?
    a Argentina **b** Russia **c** Spain **d** United Kingdom *page 106*

12 Which country is the biggest tourist destination?
    a United States **b** China **c** France **d** Italy   *page 107*

13 Which country earns the most from tourism?
    a France **b** United States **c** China **d** Italy   *page 107*

Glossary

Balance of payments The record of a country's transactions with the rest of the world. The **current account** of the balance of payments consists of: visible trade (goods); "invisible" trade (services and income); private transfer payments (eg, remittances from those working abroad); official transfers (eg, payments to international organisations, famine relief). Visible imports and exports are normally compiled on rather different definitions to those used in the trade statistics (shown in principal imports and exports) and therefore the statistics do not match. The **capital account** consists of long- and short-term transactions relating to a country's assets and liabilities (eg, loans and borrowings). The **current and capital accounts**, plus an errors and omissions item, make up the **overall balance**. **Changes in reserves** include gold at market prices and are shown without the practice often followed in balance of payments presentations of reversing the sign.

Big Mac index A light-hearted way of looking at exchange rates. If the dollar price of a burger at McDonald's in any country is higher than the price in the United States, converting at market exchange rates, then that country's currency could be thought to be over-valued against the dollar and vice versa.

CFA Communauté Financière Africaine. Its members, most of the francophone African nations, share a common currency, the CFA franc, pegged to the euro.

Cif/fob Measures of the value of merchandise trade. Imports include the cost of "carriage, insurance and freight" (cif) from the exporting country to the importing. The value of exports does not include these elements and is recorded "free on board" (fob). Balance of payments statistics are generally adjusted so that both exports and imports are shown fob; the cif elements are included in invisibles.

CIS is the Commonwealth of Independent States, including Georgia, Turkmenistan and Ukraine.

Crude birth rate The number of live births in a year per 1,000 population. The crude rate will automatically be relatively high if a large proportion of the population is of childbearing age.

Crude death rate The number of deaths in a year per 1,000 population. Also affected by the population's age structure.

Debt, foreign Financial obligations owed by a country to the rest of the world and repayable in foreign currency. The **debt service ratio** is debt service (principal repayments plus interest payments) expressed as a percentage of the country's earnings from exports of goods and services.

Debt, household All liabilities that require payment of interest or principal in the future.

Economic Freedom Index The ranking includes data on labour and business freedom as well as trade policy, taxation, monetary policy, the banking system, foreign-investment rules, property rights, government spending, regulation policy, the level of corruption and the extent of wage and price controls.

Effective exchange rate The nominal index measures a currency's depreciation (figures below 100) or appreciation (figures over 100) from a base date against a trade-weighted basket of the currencies of the country's main trading partners. The real effective exchange rate reflects adjustments for relative movements in prices or costs.

EU European Union. Members as at mid 2019 are: Austria, Belgium, Bulgaria, Croatia, Cyprus, Czech Republic, Denmark, Estonia, Finland, France, Germany, Greece, Hungary, Ireland, Italy, Latvia, Lithuania, Luxembourg, Malta, Netherlands, Poland, Portugal, Romania, Slovakia, Slovenia, Spain, Sweden and the United Kingdom (stopped being a member on January 31st 2020).

Euro area The 19 euro area members of the EU are Austria, Belgium, Cyprus, Estonia, Finland, France, Germany,

Greece, Ireland, Italy, Latvia, Lithuania, Luxembourg, Malta, Netherlands, Portugal, Slovakia, Slovenia and Spain. Their common currency is the euro.

Fertility rate The average number of children born to a woman who completes her childbearing years.

G7 Group of seven countries: United States, Japan, Germany, United Kingdom, France, Italy and Canada.

GDP Gross domestic product. The sum of all output produced by economic activity within a country. GNP (gross national product) and GNI (gross national income) include net income from abroad, eg, rent, profits.

Import cover The number of months of imports covered by reserves, ie, reserves ÷ $\frac{1}{12}$ annual imports (visibles and invisibles).

Inflation The annual rate at which prices are increasing. The most common measure and the one shown here is the increase in the consumer price index.

Life expectancy The average length of time a baby born today can expect to live.

Literacy is defined by UNESCO as the ability to read and write a simple sentence, but definitions can vary from country to country.

Median age Divides the age distribution into two halves. Half of the population is above and half below the median age.

Money supply A measure of the "money" available to buy goods and services. Various definitions exist. The measures shown here are based on definitions used by the IMF and may differ from measures used nationally. Narrow money (M1) consists of cash in circulation and demand deposits (bank deposits that can be withdrawn on demand). "Quasi-money" (time, savings and foreign currency deposits) is added to this to create broad money.

OECD Organisation for Economic Co-operation and Development. The "rich countries" club was established in 1961 to promote economic growth and the expansion of world trade. It is based in Paris and now has 36 members from May 2018, when Lithuania joined.

Official reserves The stock of gold and foreign currency held by a country to finance any calls that may be made for the settlement of foreign debt.

Opec Set up in 1960 and based in Vienna, Opec is mainly concerned with oil pricing and production issues. The current members (2019) are: Algeria, Angola, Congo-Brazzaville, Ecuador (until January 2020), Equatorial Guinea, Gabon, Iran, Iraq, Kuwait, Libya, Nigeria, Saudi Arabia, United Arab Emirates and Venezuela.

PPP Purchasing power parity. PPP statistics adjust for cost of living differences by replacing normal exchange rates with rates designed to equalise the prices of a standard "basket" of goods and services. These are used to obtain PPP estimates of GDP per head. PPP estimates are shown on an index, taking the United States as 100.

Real terms Figures adjusted to exclude the effect of inflation.

SDR Special drawing right. The reserve currency, introduced by the IMF in 1970, was intended to replace gold and national currencies in settling international transactions. The IMF uses SDRs for book-keeping purposes and issues them to member countries. Their value is based on a basket of the US dollar (with a weight of 41.73%), the euro (30.93%), the Chinese renminbi (10.92%), the Japanese yen (8.33%), and the pound sterling (8.09%).

List of countries

	Population	GDP	GDP per head	Area	Median age
	m, 2018	$bn, 2018	$PPP, 2018	'000 sq km	yrs, 2018
Afghanistan	37.2	19	1,955	653	17.9
Albania	2.9	15	13,364	29	35.8
Algeria	42.2	174	15,482	2,382	28.1
Andorra	0.1	3	...	0.5	44.9
Angola	30.8	106	6,452	1,247	16.6
Argentina	44.4	520	20,611	2,780	31.1
Armenia	3.0	12	10,343	30	34.8
Australia	24.9	1,434	51,663	7,741	37.6
Austria	8.9	455	55,455	84	43.4
Azerbaijan	10.0	47	18,044	87	31.5
Bahamas	0.4	12	32,088	14	31.7
Bahrain	1.6	38	47,303	1	32.0
Bangladesh	161.4	274	4,372	148	26.8
Barbados	0.3	5	18,554	0.4	39.9
Belarus	9.5	60	19,995	208	40.0
Belgium	11.5	543	51,408	31	41.7
Benin	11.5	10	2,425	115	18.6
Bermuda	0.1	7	52,547	4	43.5
Bolivia	11.4	40	7,873	1,099	25.0
Bosnia & Herz.	3.3	20	14,624	51	42.1
Botswana	2.3	19	18,616	582	23.6
Brazil	209.5	1,869	16,096	8,516	32.7
Brunei	0.4	14	80,920	6	31.3
Bulgaria	7.1	65	21,960	111	44.1
Burkina Faso	19.8	14	1,985	274	17.4
Burundi	11.2	3	744	28	17.2
Cambodia	16.3	25	4,361	181	25.0
Cameroon	25.2	39	3,785	475	18.5
Canada	37.1	1,713	48,130	9,985	40.8
Cayman Islands	0.1	6	72,608	0.3	40.2
Central African Rep.	4.7	2	860	623	17.4
Channel Islands	0.2	11	...	0.2	42.2
Chad	15.5	11	1,968	1,284	16.4
Chile	18.7	298	25,223	757	34.7
China	1427.6	13,608	18,237	9,563	37.7
Colombia	49.7	331	15,013	1,142	30.6
Congo-Brazzaville	5.2	11	5,662	342	19.1
Congo-Kinshasa	84.1	47	932	2,345	16.9
Costa Rica	5.0	60	17,671	51	32.6
Croatia	4.2	61	27,580	57	43.8
Cuba	11.3	100	12,300	110	41.6
Curaçao	0.2	3	27,743	0.4	41.3
Cyprus	1.2	25	38,513	9	36.3
Czech Republic	10.7	245	39,744	79	42.5
Denmark	5.8	356	55,671	43	42.0
Dominican Rep.	10.6	86	17,748	49	27.4
Ecuador	17.1	108	11,734	256	27.3
Egypt	98.4	251	12,412	1,001	24.5

	Population	GDP	GDP per head	Area '000 sq	Median age
	m, 2018	$bn, 2018	$PPP, 2018	km	yrs, 2018
El Salvador	6.4	26	8,332	21	26.8
Equatorial Guinea	1.3	13	22,744	28	22.2
Eritrea	3.5	2	1,028	118	19.2
Estonia	1.3	31	35,974	45	42.1
Eswatini	1.1	5	10,638	17	20.3
Ethiopia	109.2	84	2,022	1,104	19.0
Fiji	0.9	6	10,879	18	27.5
Finland	5.5	277	48,417	338	42.9
France	65.0	2,778	45,342	549	41.9
French Guiana	0.3	...	...	84	25.0
French Polynesia	0.3	3	...	4	32.8
Gabon	2.1	17	17,876	268	22.4
Gambia, The	2.3	2	2,612	11	17.7
Georgia	4.0	18	12,005	70	38.1
Germany	83.1	3,948	53,075	358	45.8
Ghana	29.8	66	4,747	239	21.2
Greece	10.5	218	29,592	132	44.7
Guadeloupe	0.4	...	...	2	43.0
Guam	0.2	6	35,600	1	30.9
Guatemala	17.2	78	8,462	109	22.3
Guinea	12.4	11	2,505	246	17.6
Guinea-Bissau	1.9	1	1,799	36	18.6
Guyana	0.8	4	8,641	215	26.0
Haiti	11.1	10	1,867	28	23.5
Honduras	9.6	24	5,139	112	23.6
Hong Kong	7.4	363	64,597	1	44.2
Hungary	9.7	158	31,103	93	42.7
Iceland	0.4	26	57,303	103	36.9
India	1352.6	2,719	7,763	3,287	27.8
Indonesia	267.7	1,042	13,080	1,914	29.2
Iran	81.8	446	19,377	1,745	31.1
Iraq	38.4	224	17,436	435	20.6
Ireland	4.8	382	83,203	70	37.5
Isle of Man	0.1	7	...	1	44.4
Israel	8.4	371	39,919	22	30.4
Italy	60.6	2,084	41,830	301	46.5
Ivory Coast	25.1	43	4,207	322	18.7
Jamaica	2.9	16	9,327	11	30.0
Japan	127.2	4,971	42,797	378	47.6
Jordan	10.0	42	9,479	89	23.1
Kazakhstan	18.3	179	27,880	2,725	30.2
Kenya	51.4	88	3,468	580	19.6
Kosovo	1.9	8	11,348	11	29.6
Kuwait	4.1	141	72,898	18	35.5
Kyrgyzstan	6.3	8	3,885	200	25.6
Laos	7.1	18	7,440	237	23.8
Latvia	1.9	34	30,305	64	43.4
Lebanon	6.9	57	13,081	10	28.8

	Population	GDP	GDP per head	Area '000 sq	Median age
	m, 2018	$bn, 2018	$PPP, 2018	km	yrs, 2018
Lesotho	2.1	3	3,219	30	23.6
Liberia	4.8	3	1,309	111	19.2
Libya	6.7	48	20,764	1,760	28.1
Liechtenstein	0.0	6	139,100	0.2	43.4
Lithuania	2.8	53	35,461	65	44.1
Luxembourg	0.6	71	113,337	3	39.5
Macau	0.6	55	123,892	0	38.6
Macedonia	2.1	13	16,359	26	38.4
Madagascar	26.3	14	1,891	587	19.2
Malawi	18.1	7	1,311	118	17.7
Malaysia	31.5	359	31,782	330	29.5
Maldives	0.5	5	15,308	0.3	29.4
Mali	19.1	17	2,317	1,240	16.2
Malta	0.5	15	42,581	0	42.0
Martinique	0.4	...	...	1	46.2
Mauritania	4.4	5	4,151	1,031	19.9
Mauritius	1.3	14	23,751	2	36.7
Mexico	126.2	1,221	19,845	1,964	28.6
Moldova	4.1	11	7,272	34	36.8
Monaco	0.0	7	115,700	0	53.8
Mongolia	3.2	13	13,800	1,564	27.8
Montenegro	0.6	6	20,690	14	38.4
Morocco	36.0	118	8,612	447	28.9
Mozambique	29.5	15	1,460	786	17.4
Myanmar	53.7	71	6,674	677	28.4
Namibia	2.4	15	11,102	824	21.5
Nepal	28.1	29	3,090	147	23.6
Netherlands	17.1	914	56,329	42	42.8
New Caledonia	0.3	11	31,100	19	33.0
New Zealand	4.7	205	41,005	268	37.7
Nicaragua	6.5	13	5,534	130	25.7
Niger	22.4	9	1,063	1,267	15.1
Nigeria	195.9	397	5,991	924	18.0
North Korea	25.6	32	...	121	34.8
Norway	5.3	434	65,511	625	39.6
Oman	4.8	79	41,860	310	30.0
Pakistan	212.2	315	5,567	796	22.4
Panama	4.2	65	25,554	75	29.2
Papua New Guinea	8.6	23	4,336	463	22.0
Paraguay	7.0	40	13,600	407	25.7
Peru	32.0	222	14,418	1,285	29.6
Philippines	106.7	331	8,951	300	25.1
Poland	37.9	586	31,337	313	40.9
Portugal	10.3	241	33,415	92	45.3
Puerto Rico	3.0	101	39,541	9	41.9
Qatar	2.8	191	126,898	12	32.0
Réunion	0.9	...	...	3	35.3
Romania	19.5	240	28,206	238	42.4

	Population m, 2018	GDP $bn, 2018	GDP per head $PPP, 2018	Area '000 sq km	Median age yrs, 2018
Russia	145.7	1,658	27,588	17,098	39.2
Rwanda	12.3	10	2,252	26	19.8
Saudi Arabia	33.7	787	55,336	2,150	31.1
Senegal	15.9	24	3,783	197	18.3
Serbia	8.8	51	17,435	88	41.2
Sierra Leone	7.7	4	1,602	72	19.1
Singapore	5.8	364	101,532	1	41.2
Slovakia	5.5	106	33,736	49	40.4
Slovenia	2.1	54	38,049	21	43.9
Somalia	15.0	5	...	638	16.5
South Africa	57.8	368	13,687	1,219	27.1
South Korea	51.2	1,619	40,112	100	42.5
South Sudan	11.0	5	1,504	664	18.8
Spain	46.7	1,419	39,715	506	43.9
Sri Lanka	21.2	89	13,474	66	33.3
Sudan	41.8	41	4,768	1,879	19.4
Suriname	0.6	4	15,510	164	28.6
Sweden	10.0	556	53,209	447	41.0
Switzerland	8.5	705	68,061	41	42.7
Syria	16.9	60	6,375	185	24.8
Taiwan	23.7	590	53,074	36	41.4
Tajikistan	9.1	8	3,450	141	22.2
Tanzania	56.3	58	3,240	947	17.8
Thailand	69.4	505	19,051	513	39.2
Timor-Leste	1.3	3	7,658	15	20.3
Togo	7.9	5	1,774	57	19.2
Trinidad & Tobago	1.4	24	32,015	5	35.2
Tunisia	11.6	40	12,503	164	32.2
Turkey	82.3	771	28,069	785	30.9
Turkmenistan	5.9	41	19,304	488	26.4
Uganda	42.7	27	2,038	242	16.4
Ukraine	44.2	131	9,249	604	40.7
United Arab Emirates	9.6	414	75,075	84	32.6
United Kingdom	67.1	2,855	45,974	244	40.3
United States	327.1	20,544	62,795	9,832	38.0
Uruguay	3.4	60	23,572	176	35.5
Uzbekistan	32.5	50	8,556	447	27.2
Venezuela	28.9	98	10,798	912	28.7
Vietnam	95.5	245	7,448	331	31.7
Virgin Islands (US)	0.1	4	...	0	42.1
West Bank & Gaza	4.9	15	5,158	6	20.3
Yemen	28.5	27	2,575	528	19.8
Zambia	17.4	27	4,224	753	17.2
Zimbabwe	14.4	31	3,030	391	18.6
Euro area (19)	339.8	13,639	40,412	2,760	44.0
World	7,631.1	84,930	17,912	132,025	30.6

Sources

AFM Research
Airports Council International, *Worldwide Airport Traffic Report*

Bank of East Asia
Bloomberg
BP, *Statistical Review of World Energy*

Cable
CAF, *The World Giving Index*
CBRE, *Global Prime Office Occupancy Costs*
Central banks
Central Intelligence Agency, *The World Factbook*
Company reports
Cornell University
Council of Tall Buildings and Urban Habitat
Credit Suisse

The Economist, www.economist.com
Economist Intelligence Unit, *Cost of Living Survey; Country Forecasts; Country Reports; Liveability Index*
Encyclopaedia Britannica
Eurostat, *Statistics in Focus*

FIFA
Finance ministries
Food and Agriculture Organisation

Global Democracy Ranking
Global Entrepreneurship Monitor
Global Internal Displacement Database
Global Terrorism Database, University of Maryland
Government statistics

H2 Gambling Capital
The Heritage Foundation, *Index of Economic Freedom*

Holman Fenwick Willan

IFPI
IMD, *World Competitiveness Yearbook*
IMF, *International Financial Statistics; World Economic Outlook*
INSEAD
Institute for Criminal Policy Research
International Civil Aviation Organisation
International Cocoa Organisation, *Quarterly Bulletin of Cocoa Statistics*
International Coffee Organisation
International Cotton Advisory Committee, *March Bulletin*
International Cricket Council
International Diabetes Federation, *Diabetes Atlas*
International Grains Council
International Institute for Strategic Studies, *Military Balance*
International Labour Organisation
International Olympic Committee
International Organisation of Motor Vehicle Manufacturers
International Publishers Association
International Rubber Study Group, *Rubber Statistical Bulletin*
International Sugar Organisation, *Statistical Bulletin*
International Telecommunication Union, *ITU Indicators*
International Union of Railways
Inter-Parliamentary Union

Johnson Matthey

McDonald's

National Institute of Statistics and Economic Studies

National statistics offices
Nobel Foundation

OECD, *Development Assistance Committee Report; Economic Outlook; Government at a Glance; OECD.Stat; Revenue Statistics*

Pew Research Centre, *The Global Religious Landscape*
Progressive Media

Reporters Without Borders, *Press Freedom Index*

Sovereign Wealth Fund Institute
Space Launch Report
Stockholm International Peace Research Institute
Swiss Re

Taiwan Statistical Data Book
The Times, *Atlas of the World*
Thomson Reuters
Trade Data & Analysis

Union of Concerned Scientists
UN, *Demographic Yearbook; National Accounts; State of World Population Report; World Fertility Report*
UNAIDS
UNCTAD, *Review of Maritime Transport; World Investment Report*
UNCTAD/WTO International Trade Centre
UN Development Programme, *Human Development Report*

UNESCO Institute for Statistics
UN High Commissioner for Refugees
UN Office on Drugs and Crime
UN, Population Division
US Department of Agriculture
US Energy Information Administration
US Federal Aviation Administration
US Geological Survey

Visionofhumanity.org

Walk Free Foundation
WHO, *Global Health Observatory; Global Immunisation Data; World Health Statistics*
World Bank, *Doing Business; Global Development Finance; Migration and Remittances Data; World Development Indicators; World Development Report*
World Bureau of Metal Statistics, *World Metal Statistics*
World Economic Forum, *Global Competitiveness Report*
World Federation of Exchanges
World Health Organisation
World Intellectual Property Organization
World Tourism Organisation, *Yearbook of Tourism Statistics*
World Trade Organisation, *Annual Report*

Yale University

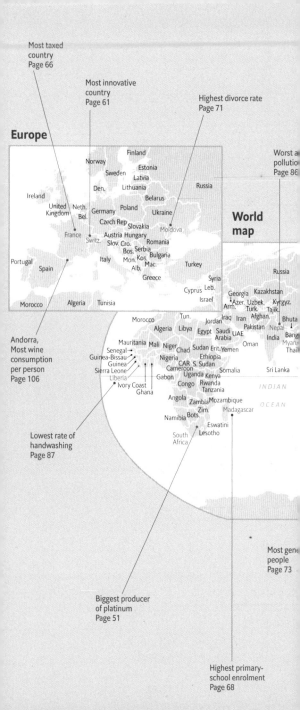